ARCHAEAL SECRETS UNVEILED: APPLICATIONS OF HALOALKALIPHILIC ARCHAEA IN BIOMEDICINE AND ENERGY

i

FROM AND TO ALLAH

Authors

Dr. Ghada E. Hegazy is a researcher at Marine Microbiology Department, National Institute of Oceanography and Fisheries (NIOF), Egypt. Her research focuses on the isolation of halophilic archaea and study their applications in different aspects. Dr. Hegazy has an outstanding experience in Microbial Fuel cell synthesis for energy production, using marine archaea. She is interested in the nanoparticles biosynthesis and their applications in biosensor field, molecular technique, biosorption of heavy metals and bioremediation of organic contaminants in the marine environment, and finally, she succeeds in the production of carotenoids and bio-surfactants from archaea with anticancer and antiviral activity against HCV, HBV, ADV, HSV and SARS viruses.

Dr. Tarek H. Taha is an Associate Professor at the Environmental Biotechnology Department, Genetic Engineering and Biotechnology Research Institute (GEBRI), City of Scientific Research and Technological Applications (SRTA-CITY), Alexandria, Egypt. He was a visiting Professor at Newcastle University, UK. He has his expertise in the field of Environmental Biotechnology. His research interest is concerned by the Biomonitoring and Bioremediation of environmental contaminants. He is also interested in the biosynthesis of nanoparticles and their applications in biosensors and other environmental fields, and has a great passion with Bioinformatics, Molecular techniques, and Genetic engineering. In addition, he is interested in the production of biofuel from environmental wastes, and finally, the bioconversion of environmental wastes into industrial and pharmaceutical products.

Dr. Abou-elela G. M. is a professor at Marine Microbiology Department, National Institute of Oceanography and Fisheries (NIOF), Egypt, she worked on the marine actinomycetes and their uses in many and different fields. Also she contributes in many researches related to marine archaea.

Prof. Nadia A. Soliman is a professor at Bioprocess Development Department, Genetic Engineering and Biotechnology Research Institute (GEBRI), City of Scientific Research and Technological Applications, Egypt. Her research focuses on the molecular biology, genetic engineering and microbial enzymes. Also her work interested in many aspects in bioprocess development and bio nanotechnology. She has very good experience in applying statistical experimental design for optimization, carrying out fermentation experiments, in addition to cloning and overexpression of genes for the purpose of production and scaling up processes.

Prof. Hanan Ghozlan is a professor at Microbiology and Botany Department, Faculty of Science, Alexandria University, Egypt. Prof Ghozlan interested in biotechnology, bioremediation, heavy metals toxicity, heavy metal accumulation, hepatitis C virus. Also her research focuses on extremophiles microorganisms, their habitats and their potential applications in different aspects. Also she has different studies in molecular biology and genes cloning.

Prof. Soraya A. Sabry is a professor at Microbiology and Botany Department, Faculty of Science, Alexandria University, Egypt. Prof Sabry interested in biotechnology, extremophiles, biofilm formation, biosurfactants production, microbial pigments, bioluminescence marine bacteria and halophilic archaea. Also her research focuses on biomedical applications of microbial byproducts. Also she interested in desalination, water treatment, water purification and nanotechnology.

Prof. Yasser R. Abdel-Fattah, Deputy Minister of Scientific Research, is a professor at Bioprocess Development Department, Genetic Engineering and Biotechnology Research Institute (GEBRI), City of Scientific Research and Technological Applications, Egypt. He is a microbiologist by training, who is well known in the bioprocess development sector with his valuable career and innovations. He has developed a scientific career in adopting numerical modeling for optimization of bioprocesses in Egypt, besides a good research in the bioproducts of extremophiles, namely extreme enzymes, biosurfactants, and microbial pigments. In addition, his interventions and inputs to science and technology policy making have positive impacts on the paradigm shift to use scientific research as a driving force for fostering national economy.

Contents

Contents

Contents

Contents

Contents

Contents

Abbreviations

Abbreviation	Item
μg	: Microgram
μl	: Microlitre
μm	: Micrometer
μ mole / g	: Micro mole per gram
AAE	: Antioxidant Equivalent Activity
ATCC	: American Type Culture Collection
Au	: Absolute Activity unit
BHT	: Butylhydroxytoluene
Blast	: Basic Local Alignment Search tool
Bp	: Base pair
BSA	: Bovin Serum Albumin
Caco – 2	: Colon cancer line
CAP-CTM	: Cobase Tag Man analyzer
DMEM	: Dulbecco's modified Eagle medium
DMSO	: Dimethyl sulfoxide
DNA	: Deoxyribonucleic acid
dNTP	: Deoxynucleotidetriphosphate
DW	: Distilled water
E24	: Emulsification activity after 24 hours
EC_{100}	: Safe dose
Em	: Emission
Ex	: Excitation
FBS	: Fetal Bovin Serum
FL2	: Thephycoerythrin emission Signal detector
FTIR	: Fourier-transform infrared spectroscopy
FU	: 5-fluorouracil
GC-MS	: Gas chromatography–mass spectrometry
H	: Hour
HBV	: Hepatities B virus
HCV	: Hepatities C virus
Hela	: Cervical cancer line
Hep G – 2	: Liver cancer line
I	: Current
IC_{50}	: The half maximal inhibitory concentration
IgE	: Immunoglobulin E
M	: Molarity
mA	: milli ampere
MCF – 7	: Breast cancer line
MEL	: Melanoma

Abbreviations

MFC	: Microbial Fuel Cell
mM	: Millimolar
mN/m	: Millinewton per meter
MTT	: 3-(4,5-dimethylthiazol-2-yl)-2,5-diphenyltetrazolium bromide NF
mV	: milli Volt
mW	: milli Watt
NCTC	: The National Collection of Type Culture
ND	: Not Detected
nm	: Nanometer
OD	: Optical Density
P	: Power
PBMCs	: Peripheral blood mononuclear cells
PBD	: Plackett-Burman Design
PBS	: Phosphate buffered saline
PI	: Propidium iodide
pM	: Picomole
PP	: Phenolphthalene
rDNA	: Ribosomal DNA
R	: Resistor
RNA	: Ribonucleic acid
rpm	: Round per minute
RPMI 1640	: Roswell Park Memorial Institute cell line
rRNA	: Ribosomal RNA
RT	: Retention time
S	: Surface Area
SEM	: Scanning Electronic Microscope
SI	: Selectivity index
ST	: Surface tension
TE	: Tris-EDTA buffer
V	: Volt
Wi – 38	: Normal human lung fibroblast cells
Xg	: Units of times gravity

List of used Chemicals and their suppliers

Item	Supplier	Item	Supplier
2.6-Di-tetra-butyl-p-cresol (butylatedhydroxytoluene)	ADWIC, Egypt.	Sodium bicarbonate	ADWIC, Egypt
Agar Agar	ROKO, Spain	Dipotassium hydrogen phosphate	ADWIC, Egypt
Agarose, electrophoresis grade	MP Biomedicals, USA	Meat extract	Fisher, UK
Ammoium nitrate	Winlab, USA	Zinc sulfate heptahydrate	MERCK, USA
Ammonium chloride	Riedel-de Haën, Germany	Lithium chloride	MERCK, USA
Ammonium dihydrogen phosphate	ADWIC, Egypt	Nutrient broth	Oxoid, UK
Ammonium molybdate tetrahydrate	Oxford, USA	Sulfuric acid	ADWIC, Egypt
Ammonium sulfate heptahydrate	MERCK, USA	DMEM	Lonza, USA
Annexin v. biotin	Sigma, USA	Streptavidine-fluorescein	Sigma, USA
Beef extract	Fisher, UK	Sodium carbonate	ADWIC, Egypt
Bovin Serum Albumin (BSA)	Lab M	Soluble starch	Oxford, USA
Casamino acids	AmiResco, USA	Sodium chloride	MP Biomedicals, USA
Casein, vitamin-free	Applichem, USA	Sodium dodecyl sulfate	Riedel-de Haën, Germany
Copper sulfate tetrahydrate	MERCK, USA	Phosphoric acid	ADWIC, Egypt
Citric acid, trisodium salt dihydrate (reagent grade)	AmiResco, USA	Sodium thiosulfate	ADWIC, Egypt
Cobalt(II) choride hexahydrate	Winlab, USA	Coppric chloride hydrate	Fluka chemika, Germany
Dimethyl sulfoxide	ADWIC, Egypt	Potassium nitrate	ADWIC, Egypt
Dipotassium phosphate	ADWIC, Egypt	Bromothymol blue	Daejung, SK
DNA ladder, 10 kb	Fermentas, USA	Taq polymerase	Fermentas, USA
dNTPs mix, 10 mM each	Thermo Scientific,USA	Hydroxyl methyl amino methane	Sigma, USA
EDTA disodium salt dehydrate	PSPARK, USA	Ethidium bromide	Fisher, UK
Ethanol	Adwic, Egypt	Ferrous sulfate	Aldrich, USA

FBS	Lonza, USA	Acridine orange dyes	Sigma, USA
Gelatin	Sigma, USA	Iodine	ADWIC, Egypt
Sodium potassium tartrate	Fisher, UK	Vanillin	Fisher, UK
Glucose	Fisher, UK	Acetone	ADWIC, Egypt
Glycerol	ADWIC, Egypt	Yeast extract	Biobasic Canadian INC
Hydrochloric acid	ADWIC, Egypt	X-gal	Bio Basic Canada Inc.
Hydrogen peroxide	ADWIC, Egypt	DNAse	Himedia, USA
Lactose	MERCK, USA	Manganese chloride	Winlab, USA
Magnesium chloride tetrahydrate	ADWIC, Egypt	Boric acid	ADWIC, Egypt
Manganese II sulphate	Winlab, USA	Potassium chloride	MERCK, USA
Methanol	Fisher, UK	Uric acid	MERCK, USA
MTT	Lonza, USA	Propidium iodide	Sigma, USA
Nickel choride hexahydrate	Sigma, USA	Magnesium sulfate heptahydrate	Riedel-de Haën, Germany
Nutrient agar	Oxoid, UK	Trisodium phosphate	Pharmacien, UK
Potassium nitate	Winlab, USA	Loading dye	RO6H Fermentas Life Sciences, France
RPMI 1640	Lonza. USA	Trypan blue	Lonza, USA
Sodium citrate	Riedel-de Haën, Germany	Sodium hydroxide	ADWIC, Egypt
Sodium molybdat dehydrate	MERCK, USA	Potassium dihydrogen phosphate	ADWIC, Egypt
Tetra methyl-p-phenylenediamine	Sigma, USA	Tryptophane	Sigma, USA
Trepton	Himedia, USA	Peptone	Fisher, UK
Tributyrin	Fluka, USA	Urea	ADWIC, USA
Zinc chloride	Aldrich, USA	L.glutamine	Lonza, USA
Phenol	Fisher, UK		

Culture media

All media used in this study were prepared using distilled water. pH was adjusted 11 using Na_2CO_3 and sterilized by autoclaving at 120°C for 20 min. Different media were used throughout the work. Media M1 to M6 were used for isolation of archaea. The used media had the following composition (g/l):

Media used for the isolation

- **M1:** Casamino acids, 7.5; yeast extract, 10; tri-sodium citrate, 3; KCl, 2; $MgSO_4.7H_2O$, 0.123; NaCl, 250; Na_2CO_3, 10 (**Hacěne** *et al.*, **2004**).

- **M2:** Casamino acids, 7.5; yeast extract, 10; trisodium citrate, 3; $MgSO_4 .7H_2O$, 0.3; KCl, 2; $FeSO_4.7H_2O$, 2×10^{-5}; $MnCl_2.4H_2O$, 3.6×10^{-4}; NaCl, 250; Na_2CO_3, 18 (**Xu** *et al.*, **1999**).

- **M3:** Casamino acids, 7.5; yeast extract, 10; trisodium citrate, 3; $MgSO_4.7H_2O$, 2; KCl, 10; $FeSO_4.7H_2O$, 2×10^{-5}; $MnCl_2.4H_2O$, 3.6×10^{-4}; NaCl, 250; Na_2CO_3, 10 (**Hu** *et al.*, **2008**).

- **M4**: Casamino acids, 7.5; yeast extract, 10; trisodium citrate, 3; $MgSO_4.7H_2O$, 1; KCl, 10; LiCl, 0.1; $FeSO_4.7H_2O$, 2×10^{-5}; $MnCl_2.4H_2O$, 3.6×10^{-4}; NaCl, 250; Na_2CO_3, 10 (**Fan** *et al.*, **2004**).

- **M5:** Casamino acids, 5; KH2PO4, 1; MgSO4.7H2O, 0.2; NaCl, 200; trace metals, 1ml; Na_2CO_3, 18 (**Goh** *et al.*, **2011**).

- **M6:** Casamino acids, 7.5; yeast extract, 20; trisodium citrate, 3; $MgSO_4.7H_2O$, 1; KCl, 2; $FeSO_4.7H_2O$, 2×10^{-5}; $MnCl_2.4H_2O$, 3.6×10^{-4}; NaCl, 250; Na_2CO_3, 18.5 (**Giménez** *et al.*, **2000**).

-

Media for growth and antimicrobial activity

- **Medium for preservation of fungi**: Glucose, 10; Peptone, 5; $MgSO_4$, 0.5; K_2HPO_4, 0.5; NaCl, 0.5; pH 5 **(Atlas, 2010)**.

- **Nutrient agar for antimicrobial sensitivity:** Peptone, 5; yeast extract, 3; agar, 20 **(Atlas, 2010)**.

Media for physiological and biochemical characterization

- **Indole production medium**: Tryptophan, 20; NaCl, 200; KH_2PO_4, 1; $MgSO_4.7H_2O$, 0.2; casamino acids, 2.5; trace element solution, 1ml; Na_2CO_3, 18 **(Cheesbrough, 1985).**

- **Simmon's citrate agar:** $Na_3C_6H_5O_7$, 2; $NH_4H_2PO_4$, 1; K_2HPO_4, 1; $MgSO_4.7H_2O$, 0.2; bromothymol blue, 0.08; NaCl, 200 trace element solution, 1ml; Na_2CO_3, 18 **(Simmon's, 1926).**

- **Medium for hydrogen sulfide production:** Casein, 20; meat extract, 6.1; $(NH_4)_2 Fe (SO_4)_2.6H_2O$, 0.2; $Na_2S_2O_3$, 0.2; NaCl, 200; K_2HPO_4, 1; $MgSO_4.7H_2O$, 0.2; Na_2CO_3, 18 **(Mackie & McCatney, 1996).**

- **Medium for lactose fermentation**: Lactose, 2.5; casamino acids, 2.5; KH_2PO_4, 1; $MgSO_4.7H_2O$, 0.2; NaCl, 200; trace metals, 1ml; Na_2CO_3, 18 **(Mudryk & Odgorska, 2006).**

- **Nitrate reduction medium:** Peptone, 5; NaCl, 200; KH_2PO_4, 1; $MgSO_4.7H_2O$, 0.2; casamino acids, 2.5; KNO_3, 0.2; trace metals, 1ml; Na_2CO_3, 18 **(Williams *et al.*, 1989).**

- **Medium for casein utilization:** Casein, 5; KH_2PO_4, 1; $MgSO_4.7H_2O$, 0.2; NaCl, 200; trace metals, 1ml; Na_2CO_3, 18 **(Mudryk & Odgorska, 2006).**

- **Lipolytic activity medium (Lipase production)**: Tributyrin (1%), 2.5; casamino acids, 2.5; KH_2PO_4, 1; $MgSO_4.7H_2O$, 0.2; NaCl, 200; trace metals, 1ml; Na_2CO_3, 18 **(Mudryk & Odgorska, 2006).**

- **Modified gelatine medium**: Casamino acids, 5; NaCl, 200; KH_2PO_4, 1; $MgSO_4.7H_2O$, 0.2; trace metals, 1ml; Na_2CO_3, 18. A supplement of 5% (w/v) inter-mittently-sterilized gelatine was added to the previously autoclaved medium under aseptic conditions **(Clarke, 1953)**.

- **Medium for amylase production**: Starch, 2.5; casamino acids, 2.5; KH_2PO_4, 1; $MgSO_4.7H_2O$, 0.2; NaCl, 200; trace metals, 1ml; Na_2CO_3, 18 **(Mudryk & Odgorska, 2006)**.

- **Medium for DNase production:** DNase, 2.5; casamino acids, 2.5; KH_2PO_4, 1; $MgSO_4.7H_2O$, 0.2; NaCl, 200; trace metals, 1ml; Na_2CO_3, 18 **(Mudryk & Odgorska, 2006)**.

- **Urea broth basal medium:** Yeast extract, 0.1; KH_2PO_4, 9.1; Na_2HPO_4, 9.5; phenol red, 0.01; urea, 20; casamino acids, 2.5; NaCl, 200; $MgSO_4.7H_2O$, 0.2; trace metals, 1ml; Na_2CO_3, 18 **(Mudryk & Odgorska, 2006)**.

- For solid media, 20g agar were added.

Trace metals solution contained: $ZnSO_4.7H_2O$, 0.1; $MnCl_2.4H_2O$, 0.03; H_3BO_3, 0.3; $CoCl_2.6H_2O$, 0.2; $CuCl_2.2H_2O$, 0.01; $NiCl_2.6H_2O$, 0.02; $Na_2MoO_4.H_2O$, 0.03.

Reagents

- **Kovac's indole:** (Amyl alcohol, p-dimethyl-aminobenzaldehyde, conc HCl).

- **Phosphomolybdate reagent (Prieto *et al.*,1999):** (32mM sodium phosphate, 4mM ammonium molybdate and 0.6M sulfuric acid).

- **Iodine solution (Bragger *et al.*, 1989):** (1g of potassium iodide, 2g iodine crystal in 100ml distilled water).

- **Vanillin reagent (Christopher *et al.*, 1971):** (6g of vanillin dissolved in 1000ml of distilled water and stored in brown bottle).

- **Phospho-vanillin reagent (Christopher *et al.*, 1971):** (350ml of vanillin reagent, 50ml of distillied water, 600ml of concentrated phosphoric acid).

- **Phenol reagent (Michel *et al.*, 1955):** (80% of grade phenol).

Buffers

- **TE buffer:** 10mM Tris-HCl and 1mM EDTA
- **TEN buffer:** 40mM Tris-HCl, 1mM EDTA and 150mMNaCl
- **10x TBE buffer:** 0.9M Tris, 0.9M boric acid and 20mM EDTA

Gels and stains

Agarose gel:

- **Ethidium bromide:** 1g ethidium bromide dissolved in 100ml H_2O

- **1% Agarose gel:** 1g agarose powder dissolved in 100ml 0.5X TBE buffer, heat until complete dissolving.

Microbial Diversity of Archaeal Isolates from El-Hamra Lake: Insights into Composition, Enzymatic Activities, and Growth Characteristics

Abstract

This study presents the results of chemical analysis for water and sediment samples collected from El-Hamra Lake in Wadi El-Natrun. Additionally, microbial isolates were obtained from various sites within the lake, and their characteristics were investigated. To differentiate between archaeal and non-archaeal isolates, specific primers were used in PCR analysis. Out of the 52 isolates tested, 37 showed positive results for archaeal DNA, indicating their classification as archaea. Further qualitative tests were conducted to assess the ability of haloarchaeal isolates to produce hydrolases. The isolates were grown in a medium supplemented with different substrates specific to the tested hydrolytic enzymes. In addition to the enzymatic analysis, the archaeal isolates were evaluated for their ability to produce biosurfactants. The growth of the isolated archaea was evaluated under different conditions, including NaCl concentration, pH, and temperature. Overall, this study provides valuable insights into the chemical composition of El-Hamra Lake and ant its microbial diversity. Additionally, the identification of biosurfactant-producing archaeal isolates and their growth characteristics contributes to the understanding of potential applications in various industries.

Keywords: El-Hamra Lake, Archaeal isolation, Enzymatic analysis, Biosurfactant production.

1

Introduction

Microbial diversity plays a crucial role in maintaining ecological balance and carrying out essential biochemical processes in various ecosystems, including aquatic environments. Archaea, a distinct domain of microorganisms, have gained significant attention in recent years due to their unique physiological and biochemical characteristics. El-Hamra Lake, a high saline lake renowned for its ecological significance, offers an ideal setting to explore the microbial diversity of archaeal isolates and gain insights into their composition, enzymatic activities, and growth characteristics **(Poli et al., 2017).** El-Hamra Lake, represents an intriguing ecosystem with its diverse physical and chemical features, including temperature, salinity, and nutrient availability. Such environmental parameters have a profound impact on the microbial communities inhabiting the lake, including archaea. Archaea are prokaryotic microorganisms that thrive in extreme environments, such as high temperatures, high salinity, and low oxygen concentrations. However, recent studies have revealed their presence and significant contribution to various non-extreme environments as well, including soda lakeecosystems like El-Hamra Lake **(Mesbah et al., 2007).** Understanding the composition of archaeal communities in El-Hamra Lake is essential for comprehending the overall microbial diversity and ecological functioning of the lake. Molecular techniques, such as DNA sequencing and metagenomics, have revolutionized the identification and characterization of microbial communities, enabling researchers to delve into the microbial world with unprecedented precision. By employing these advanced techniques, scientists have been able to unravel the intricate web of archaeal diversity in various environments, shedding light on their ecological roles and metabolic capabilities. In addition to their compositional analysis, investigating the enzymatic activities of archaeal isolates from El-Hamra Lake is of great interest. Archaea possess unique enzymatic machinery that enables them to carry out specific biochemical reactions, including the production and degradation of various organic compounds. Exploring the enzymatic activities of archaeal isolates can provide valuable insights into their functional diversity and potential contributions to biogeochemical cycles in the lake ecosystem. Furthermore, an understanding of the growth characteristics of archaeal isolates from El-Hamra Lake is crucial for comprehending their adaptability and survival strategies in this specific environment. Growth rate, temperature tolerance, pH range, and nutrient requirements are among the key parameters that shape the ecological niche of microorganisms **(Shakya et al., 2013)**. By studying the growth characteristics of archaeal isolates, researchers can gain insights into their ecological preferences and potential interactions with other microbial communities within the lake. In this study, we aim to investigate the microbial diversity of archaeal isolates from El-Hamra Lake using advanced molecular techniques. We will analyze the composition of archaeal communities, explore their enzymatic activities, and characterize their growth characteristics. The findings from this study will contribute to our understanding of the ecological roles and metabolic potential of archaea in soda lake ecosystems, specifically focusing on El-Hamra Lake. Moreover, this research will provide valuable insights into the broader context of microbial diversity and ecosystem dynamics, aiding in the development of sustainable environmental management practices. Overall, this chapter serves as a pioneering effort to unravel the microbial diversity of archaeal isolates from El-Hamra Lake, shedding light on their composition, enzymatic activities, and growth characteristics. The knowledge gained from this research will contribute to our understanding of microbial ecology, biogeochemical cycles, and the intricate relationships between microorganisms and their environment in soda lakeecosystems **(Mesbah et al., 2007)**

Methods

Site description of the isolation area

El-Hamra Lake located in Wadi El-Natrun depression, Egypt, was the source of the isolation of the halophilic archaea. Its geographical coordinates are 30° 23` 21`North, and 30° 20` 45`East. It is an alkaline inland saline lake with elongated depression about 90 km northwest of Cairo **(Fig. 1)**. Its average length is about 60 km and average width about 10 km. The bottom of the Wadi El-Natrun is 23 m below sea level and 38 m below the water level of Rosetta branch of the Nile. El Hamra Lake appears to be unique among the saline lakes that were previously characterized by alkaline brine poor in Ca and Mg. The dilute HCO_3-CO_3 spring water evolves into alkaline Na-SO_4-Cl-rich brine **(Abd-el-Malek & Rizk, 1963)**.

Fig. 1: A map showing the location of El-Hamra Lake in Wadi El-Natrun.

Chemical analysis of El-Hamra Lake samples

The chemical analysis of water and sediments of El-Hamra Lake, Wadi El-Natrun, was carried out at the central Lab, Scientific Researches and Technological Applications City, Borg El-Arab, Alexandria, Egypt.

Isolation and selection of haloalkaliphilic archaea

For the enrichment of halophilic archaea, 6 different isolation broth media named M1, M2, M3, M4, M5, and M6 (Annex II) were used. One ml of water or 1g sediment was used to inoculate100 ml of the isolation medium in 250 ml Erlenmeyer flasks. Flasks were incubated on a rotary shaker at 200 rpm at 37°C for 7days. After incubation, 1ml from each flask (serially diluted) was transferred into a Petri plate containing the same agar medium and cultivated under the same conditions **(Salgaonkar et al., 2013)**. The obtained isolates were purified by streaking on the same isolation agar media to ensure purity. Pure separate colonies were maintained on slants with the same isolation media and were stored at 4°C and sub-cultured regularly.

Molecular Identification of the isolates

To select archaea isolates, a molecular approach, based on archaeal 16S rDNA genes, was employed.

- Chromosomal DNA preparation

A simplified rapid protocol for preparing DNA from bacterial isolates was used in this study. This protocol is an extension of a method described originally for the preparation of plasmid DNA (Holmes & Quigley, 1981). The principle depends on a rapid disruption of cells from individual colonies picked from an agar medium (Moore et al., 2004). Individual colonies from an agar plate were picked (1–5 colonies are usually adequate for generating sufficient DNA) using a sterile toothpick and resuspended in 100 µl sterile TE Buffer (Annex III). The cell suspension was placed in a water bath at 97°C for 10 min. The cell lysate was centrifuged (13,000 rpm, 5 min), and the supernatant containing DNA was removed, and an aliquot (1–5 µl) was added to a PCR reagent mixture.

- Amplification of the 16S rDNA gene

Polymerase chain reaction (PCR) was carried out in order to amplify the 16S rDNA genes from archaeal genomes using universal primers designed to almost amplify the full length (~1500 bp) of this gene. The sequence of the forward primer was 5'ATTCCGGTTGATCCTGCCGG-3', while the sequence of the reverse primer was 5'-TACGGYACCTTGTTACGACT-3 as reported by **(Salgaonkar et al., 2013)** for the archaea primers. Reaction mixture contained 0.05 μg of the template DNA in a final volume of 25 μl, 0.2 mM dNTPs mix, 1X polymerase buffer, 1U Taq polymerase and 0.2 pM of each primer. Thermocycling conditions were as follows: DNA was denatured at 95°C for 4 min, then 30 cycles of denaturation (94°C for 1 min), annealing (55°C for 1 min) and extension (72°C for 2 min) were performed, and at the end of the last cycle, the extension time was lengthened by 10 min **(Salgaonkar et al., 2013)**.

- Agarose gel electrophoresis

Gel electrophoresis was carried out according to the method described by **(Sanbrook et al., 1989)**. Agarose (1%) was dissolved in 0.5X TBE buffer (Annex III). The agarose mixture was completely dissolved by heating at 100ºC, then the solution was allowed to cool to about 50ºC, after which 5 μl of 10 mg/ml of ethidium bromide-staining solution was added to the solution. The solution was poured into the electrophoresis chamber and was allowed to cool. The electrophoresis buffer (0.5X TBE buffer) was used to completely submerge the gel. The electrophoresis was carried out at a constant voltage of 80 Volts for about 45 min. The DNA was visualized and photographed using UV-transilluminator (UVP Dual intensity transilluminator).

Biochemical characterization of archaeal isolates and cluster analysis

- Indole production test

Sterile tryptophan broth (Annex II) was inoculated with the corresponding archaeal isolates and incubated at 37°C for 7 days. Kovac's reagent (0.5ml) (Annex III) was added, and then well shaken. Positive result was indicated by the appearance of a red colour in the upper reagent layer due to the presence of indole **(Cheesbrough, 1981)**.

- ### Citrate utilization test

Simmon's solid citrate medium (Annex II) was inoculated and incubated for 7days at 37°C. In the positive reaction, the incubation will be blue (with_streak of growth) due to the alkaline reaction. In the negative reaction, the slope will be green with no growth **(Simmons, 1926)**.

- ### Hydrogen sulfide production

The tested archaea were streaked aseptically on the surface of M5 medium slant (Annex II), and then incubated at 37°C for 7 days. The resultant culture over the surface of the slant was then suspended in 5 ml of 20% NaCl solution. An aliquot of 0.1 ml of NaCl suspension of each of the tested isolates was then streaked, each separately, on ferrous sulfate containing slants and incubated at 37°C for 7 days **(Collee et al., 1996)**.

Enzymatic activities (Mudryk & Odgorska, 2006)

- ### Lactose fermentation

Liquid media containing phenol red (Annex II) as indicator was inoculated with tested organisms and incubated at 37° C for 7 days. The change in color from red to yellow indicates the ability of the isolates to ferment lactose to glucose then to pyruvic acid.

- ### Nitrate reduction

KNO_3 containing liquid medium (Annex II) was inoculated with the experimental bacteria and incubated at 37°C for 7 days. Zn^{2+} was added, a negative result was indicated by the change of color to red.

- ### Caseinase production

The isolates were streaked, on M5 agar plates containing casein (1%) as the sole carbon and the nitrogen source, and were then incubated at 37°C for 7 days. The appearance of a clear zone indicates the production of enzyme as a positive result.

- ### Lipase production

The isolates were streaked on M5 agar plates containing tributyrin (1%) dissolved with gum (the substrate for lipase enzyme) and incubated at 37°C for 7days. The appearance of clear zone around the growth indicates the positive result and the production of the lipase enzyme.

- ### Gelatinase production

Bacteria were streaked on nutrient gelatine medium (Annex II). The slants were incubated for at least 7 days, and then placed into the refrigerator for approximately 30 min. If the medium solidifies in the refrigerator, then a negative test result was recorded. If the organism produces sufficient gelatinase, the tube will remain liquid, at least partially **(Clarke, 1953)**.

- ### Amylase production

The isolates were incubated at 37°C for 7 days on an agar plate containing 0.4% soluble starch (Annex II). After incubation, few drops of iodine solution (as an indicator) (Annex III) were added to the surface of the starch agar plate. If the organism produces amylase, a clear zone will surround the colonies.

- ### Urease production test

The medium was slanted and the tubes were inoculated and incubated at 37°C for 7days after incubation, phenol red was used as indicator. Positive urease production is indicated by the color change from orange yellow to bright pink.

- ### DNase production test

The isolates were incubated at 37°C for 7 days on an agar plate containing DNA (Annex II). After incubation, if the organism produces DNase enzyme, a clear zone will surround the colonies.

Cluster analysis based on biochemical tests

Computational approaches were used for clustering the archaeal isolates based on the results of their biochemical tests via creating a binary matrix. The resultant patterns were analyzed using Multi Experiment Viewer v4.9 (MeV software; available at *(http://www.tm4.org/mev.html)* (AI et al., 2003). Hierarchical clustering (HCL) was performed and dendrograms representing results obtained from biochemical tests were generated based on the Eucladian distance matrix. The resultant Newick trees were then displayed using MEGA version 4.0.2 **(Tindall et al., 1984)**.

The results of the biochemical tests were converted to a binary code matrix of (0, 1) type. For instance, positive test was given the 1 code and negative was given the 0 code. Multi Experiment Viewer v4.9 (MeV) was then used to analyze the resultant binary pattern existed in Excel-Sheet. The first step in using MeV is to import data from a text file, which must be organized as a tab-delimited field values. Such tab-delimited text files can be created and exported in any standard spread sheet program, such as Microsoft Excel. Clustering was then performed to group similar strains together.

Identification of M6 archaeal isolate

Archaeal cells were processed by Gram stain using bio-diagnostic Gram stain kit. For electron microscopy, sample was metalized with a thin gold film using sputtering device (JFC-1100 E JOEL, USA) for 12 min. Scanning electron microscopy was performed with JSM 5300 JOEL, USA Scanning Electron Microscope at 20 kV in the Central Laboratory, City of Scientific Research and Technological Applications.

- **PCR product purification**

The PCR product from previous experiment was purified to remove unincorporated nucleotides and excess primer using PCR purification kit (QIA gen PCR purification Kit). The PCR product was mixed with 5 volumes of phosphate buffer (included in the kit) and the mixture was loaded onto the column. After 30 seconds of centrifugation at 13,000 rpm, the column was washed with 0.75 ml of TE buffer and centrifuged for 1 min to dry. The DNA was eluted with 50 μl of elution buffer **(Falchetti et al., 2005)**.

- **DNA sequencing and analysis**

The purified PCR product was sequenced using dideoxy chain termination method **(Sanger et al., 1977)**. This was done using ABI PRISM model 3730 automated DNA sequencer at Sigma for Scientific Research and big dye terminator ready reaction mix. The sequencing reaction was performed with four different fluorescent labelled ddNTPs, instead of radioactive labels. The thermal cycling mixture was as following: 8 μl big dye terminator mix, 6 μl of the primer (10 pM) and 6 μl of the template. The

9

universal archaeal primers were used for fishing a partial sequence of obtained 16S-rRNA gene.

The sequences were assembled using BioEdit Sequence Alignment Program **(Hall, 1999)**, and comparative sequence analyses were performed using ClustalW modulus of the BioEdit program. BLAST program was used to assess the similarity and phylogenetic trees were constructed with MEGA software version 4.0.2 **(Tamura et al., 2007)**.

Correlation between the optical density and dry weight measurement of *Natrialba* sp. M6

In a preliminary step, a correlation curve between the growth of *Natrialb*a sp. M6 measured as OD_{600} and the dry weight was plotted. The correlation factor was determined from the linear relation between both. In this experiment, the bacterium was allowed to grow in 100 ml of M5 in 250 ml conical flask. At time intervals, OD was measured using spectrophotometer. For dry weight determination, cultures were centrifuged and cell pellets were washed twice with saline solution 0.9% (w/v) NaCl isotonic solution, centrifuged at 10000 rpm for 15 min, and dried at 105°C until constant weight **(Widdel, 2007)**.

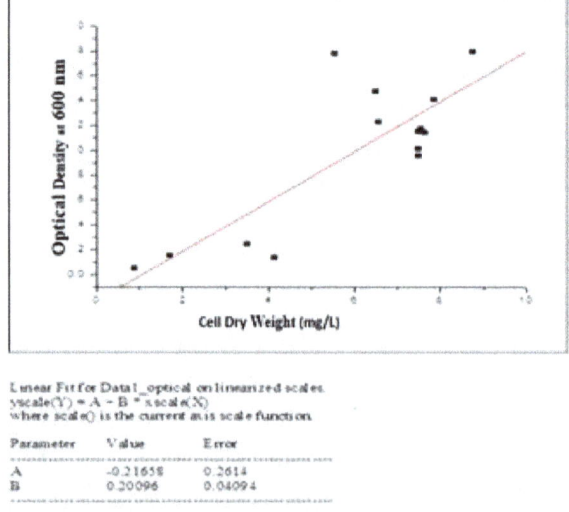

Fig 2: Correlation between optical density and dry weight measurements.

Effect of NaCl concentration, pH and temperature on the growth of *Natrialba* sp. M6

Natrialba sp. M6 was cultivated on basal medium amended with different NaCl concentrations (5, 10, 15, 20, 25, and 30%) to determine optimum concentration for growth. For determining optimum pH, the medium was adjusted at different pH values (7, 8, 9, 10, 11, and 12). Incubation at different temperatures (30, 37, 40, 45, and 50 ºC) was also examined. Cultures were incubated on shaker (200 rpm) for 7 days and growth was monitored in terms of OD_{600} (**VijayAnand et al., 2010**).

Results

Chemical analysis for water and sediment samples

Data of chemical analysis for water and sediments samples of El-Hamra Lake, Wadi El-Natrun are presented in **Tables (1 & 2)** including potassium, magnesium, calcium, silica, phosphate, chlorine, sulfate, CO_3, HCO_3 and total nitrogen.

Table 1: Water samples analysis (mg/l)

Parameter	April/2013	September /2013
Potassium	28.96	31.49
Magnesium	9.04	9.00
Calcium	2.85	1.17
Silica	270.00	270.90
Phosphate	1214.00	1202.00
Chloride	20.99×10^3	56.10×10^3
Sulfate	114697.00	115303.00
CO_3	3000.00	900.00
Total nitrogen	34.72	61.60
Total Alkalinity (HCO_3)	259200.00	242000.00

Table 2: Sediments samples analysis (mg/g)

Parameter	April /2013	September /2013
Potassium	3.10	2.05
Magnesium	0.44	2.12
Calcium	0.96	1.90

Selection of archaea and screening for biosurfactant production

- Using archaeal primers

In this study, 52 microbial isolates were recovered from different sites of water and sediments from El-Hamra Lake, Wadi El-Natrun. Most of the colonies obtained showed different shades of pink, orange, and red colors, while others had white, beige and colorless.

Archaeal specific primers were used to differentiate between archaeal and non-archaeal isolates. When all isolates were tested with PCR using the archaeal specific forward and reverse primers, 37 out of 52 gave positive results (a band of about 1,500 bp), and have been categorized as archaea and selected for the subsequent work.

- Biochemical tests of the archaeal isolates

The ability of the haloarchaeal isolates to produce hydrolases and other biochemical tests was determined qualitatively **(Table 3)**. For this objective, the growth medium M5 was supplemented with 10 different substrates separately according to the tested hydrolytic enzyme. The qualitative enzymatic detection was recorded and the results were expressed as (+) or (-). The isolates coded GH8, RE9, S2, RA4, S8, M3, and RA7 were able to produce amylase and DNase. None of the isolates was able to produce caseinase, gelatinase, citrate permease, lipase, urease, tryptophanase, β-galactosidase or hydrogen sulfide. All archaeal isolates were able to produce oxidase, while nitrate reductase was produced by all except S8. It is interesting to note that combined hydrolytic activity was also detected in many isolates. Four isolates presented 5 hydrolytic activities, 20 isolates presented 4 hydrolytic activities, 9 isolates presented 3, while 4 isolates presented 2 hydrolytic activities. The relationship between archaeal isolates according to the biochemical tests and hydrolytic enzymes is shown in **Fig. (3)** after converting the data in **Table (3)** into numerical binary matrix including the values (0) and (1) for negative (-) and (+/or more), respectively. It was recognized, the isolates were categorized into four clusters I, II, III, and IV, where each cluster was subdivided into two sub-cluster except the third cluster.

Table 3: Biochemical characterization of the haloarchaeal isolate

Isolate Codes	Oxidase	Catalase	Amylase	DNAse	Nitrate reductase	Lipase	Gelatinase	Urease	Citrate permease	Casinase	β-galactosidase	Tryptophanase	H$_2$S production
M6	++++	++	+	-	+++	-	-	-	-	-	-	-	-
A8	+++	++	-	-	+++	-	-	-	-	-	-	-	-
S4	++	+	++	-	++	-	-	-	-	-	-	-	-
A3	+	-	-	-	+	-	-	-	-	-	-	-	-
GH8	++	-	+	+	+++	-	-	-	-	-	-	-	-
RA1	+++	++	++	-	+++	-	-	-	-	-	-	-	-
A5	++++	++	+	-	+++	-	-	-	-	-	-	-	-
A7	+++	++	+	-	+++	-	-	-	-	-	-	-	-

M2	++	+	+	-	+++	-	-	-	-	-	-	-	-
RE2	++	+	-	-	+++	-	-	-	-	-	-	-	-
RE6	+++	+	-	-	++	-	-	-	-	-	-	-	-
A1	+	-	+	-	+	-	-	-	-	-	-	-	-
A6	++	+	+	-	++++	-	-	-	-	-	-	-	-
RE9	+++	-	++	+	+++	-	-	-	-	-	-	-	-
RE5	++	-	-	-	++	-	-	-	-	-	-	-	-
RE8	+++	+	++	-	++	-	-	-	-	-	-	-	-
RE10	++++	+	-	++	+	-	-	-	-	-	-	-	-
M1	++++	+	-	+	+++	-	-	-	-	-	-	-	-
S10	+	+	-	-	+	-	-	-	-	-	-	-	-
A2	+++	+	-	+	+	-	-	-	-	-	-	-	-
RE4	+++	++	-	-	++	-	-	-	-	-	-	-	-
M7	+++	+	-	-	+++	-	-	-	-	-	-	-	-
YR7	++	-	-	-	++++	-	-	-	-	-	-	-	-
S2	++	+	++	+	++++	-	-	-	-	-	-	-	-
RA4	+	+	+++	++	++	-	-	-	-	-	-	-	-
RA2	++	+	-	-	++	-	-	-	-	-	-	-	-
GH7	+++	+++	+	-	+	-	-	-	-	-	-	-	-

YR6	++	-	-	-	++	-	-	-	-	-	-	-
GH6	++	+	-	++	++	-	-	-	-	-	-	-
S8	++	+++	+++	+	-	-	-	-	-	-	-	-
M3	++	+++	+	+	++	-	-	-	-	-	-	-
RA8	+++	+++	+	-	+	-	-	-	-	-	-	-
YR8	++	+++	+	-	++	-	-	-	-	-	-	-
S9	+	+	++	-	++++	-	-	-	-	-	-	-
RA3	++++	+	-	-	++	-	-	-	-	-	-	-
RA7	+++	+	+++	+	++	-	-	-	-	-	-	-
GH1	++	+	+	-	++	-	-	-	-	-	-	-

Response was not detected (-), week (+), Moderate (++) and strong response (+++)

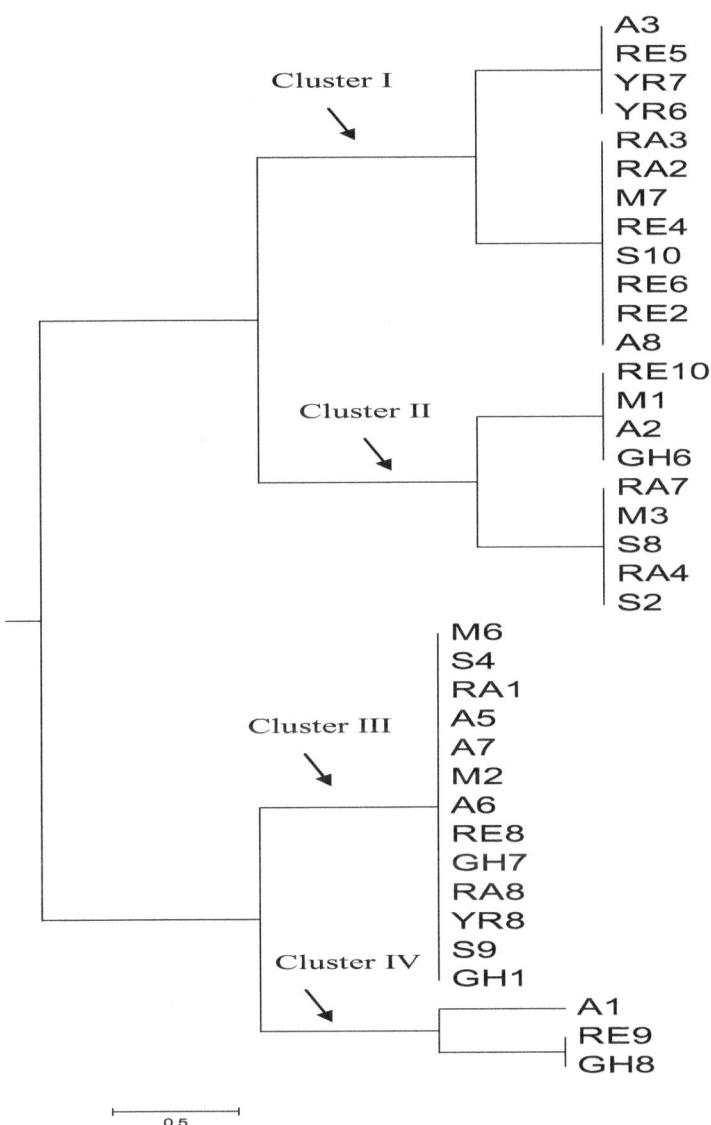

Fig. 3: Phenotypic tree showing the relationship between the halophilic archaeal isolates from El-Hamra Lake, Wadi El-Natrun according to visual detection for different hydrolytic enzymes and some biochemical tests.

Screening for biosurfactant production

In this experiment, the archaeal isolates were screened for their capability to produce biosurfactant. This was achieved through the application of four techniques: surface tension measurement, oil displacement, emulsification index (EI 24%), and blood haemolysis.

Based on the data obtained **(Table 4)**, only two strains (M6 and A5) gave significantly positive and reproducible results and were considered as biosurfactant-producers.

Table 4: Screening of biosurfactant-producing haloarchaea isolates

Isolates	Surface Tension mN/m	Haemolytic test	Oil displacement test	Emulsification Index (EI)
M6	45.0	+	+	1.5 / 2.5
A8	37.6	-	+	-
A3	48.8	-	+	-
RA1	46.2	-	+	-
A5	46.2	+	+	1.3 / 2.5
RE5	47.9	-	+	-
M7	46.6	-	+	-
RA2	44.7	-	+	-
GH6	44.5	-	+	-
S8	47.3	-	+	-
Control	65.0	-	-	-

Identification and characterization of the archaeon isolate M6

- Phenotypic characterization

The most promising biosurfactant producing archaeal isolate M6 was Gram-negative cells, small round orange pigmented colonies, when grown on solid medium (M5) **(Fig. 4A & B)**. Under SEM, the cells appeared rod-shaped ranging from 0.613, 0.925 μm to 0.441, 0.516 μm **(Fig. 4C)**.

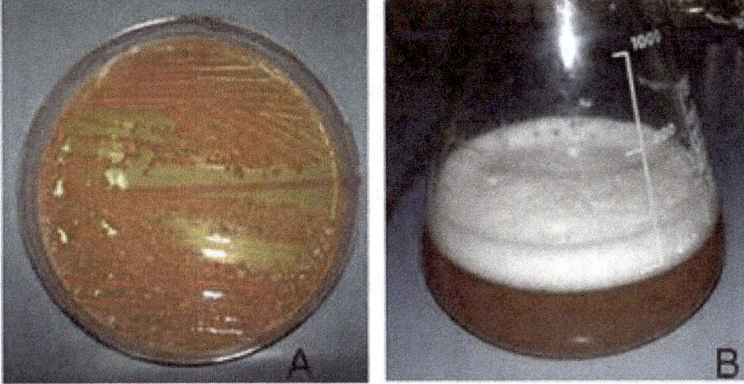

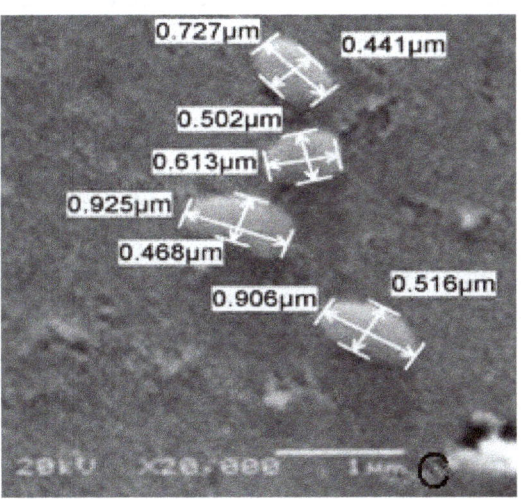

Fig. 4: Orange pigmented colonies of the archaeon M6 grown on M5-agar (A), growth on M5 broth (B) and scanning micrograph at 20000X magnification (C).

- **Amplification of 16S rRNA gene**

The archaeal isolate M6 was cultured for 7 days at 37°C on M5 and its DNA was extracted, as described previously in the method section, and used as a template for

PCR amplification. The product of the PCR was analyzed on 1% agarose gel stained with ethidium bromide. Finally, the selected isolate M6 was identified by sequencing of the PCR amplified 16S rRNA gene where it gave 1322 bp. The nucleotide sequence was searched for homology against GenBank database using Nucleotide BLAST search tool (Blastn), where it showed 99% similarity to *Natrialba chahannaoensis* strain WNHS9 with accession number (KP828441) **(Table 5)**. The obtained homology matches for the isolate was then aligned together with the isolate sequence to obtain multiple sequence alignment. The multiple sequence alignments were used to construct a phylogenetic tree using BioEdit Sequence Alignment Editor Program. The constructed phylogenetic tree was presented in **(Fig. 5).** Subsequently, the 16S rRNA of the selected isolate (M6) was submitted into GenBank under accession number (MK063890).

Table 5: 16S rRNA gene sequences with close similarity to M6 archaeal strain, accession numbers, scores, query coverage, sequence identity percentages to the nearest neighbors based on 16S rRNA sequences data.

Description	Max Score	Total Score	Query Cover	E - value	Ident	Accession
Natrialba chahannaoensis strain WNHS9 16S ribosomal RNA gene, partial sequence	2377	2377	99%	0.0	99%	KP828441.1
Natrialba chahannaoensis strain WNHS20 16S ribosomal RNA gene, partial sequence	2372	2372	99%	0.0	99%	KP828442.1
Natrialba chahannaoensis strain WNHS4 16S ribosomal RNA gene, partial sequence	2372	2372	99%	0.0	99%	KP828440.1
Natrialba chahannaoensis strain WNHS2 16S ribosomal RNA gene, partial sequence	2372	2372	99%	0.0	99%	KP788716.1
Natrialba chahannaoensis strain JCM 10990 16S ribosomal RNA gene, complete sequence	2372	2372	99%	0.0	99%	NR113520.1
Natrialba sp. strain S14 16S ribosomal RNA gene, partial sequence	2368	2368	99%	0.0	99%	MF150021.1
Natrialba chahannaoensis strain w 16S ribosomal RNA gene, partial sequence	2362	2362	99%	0.0	98%	KP676930.1
Natrialba wudunaoensis strain	2361	2361	99%	0.0	98%	EU672838.1

Sua-E42 16S ribosomal RNA gene, partial sequence						
Natrialba chahannaoensis strain Sua-CS1 16S ribosomal RNA gene, partial sequence	2361	2361	99%	0.0	98%	EU672837.1
Natrialba sp. SSLVU01 16S ribosomal RNA gene, partial sequence	2350	2350	99%	0.0	98%	KJ123850.1
Natronobacterium sp. ATCC 43988 DNA for 16S rRNA	2348	2348	99%	0.0	98%	D 88256.1
Natrialba chahannaoensis strain C112 16S ribosomal RNA gene, complete sequence	2344	2344	99%	0.0	98%	NR028181.1

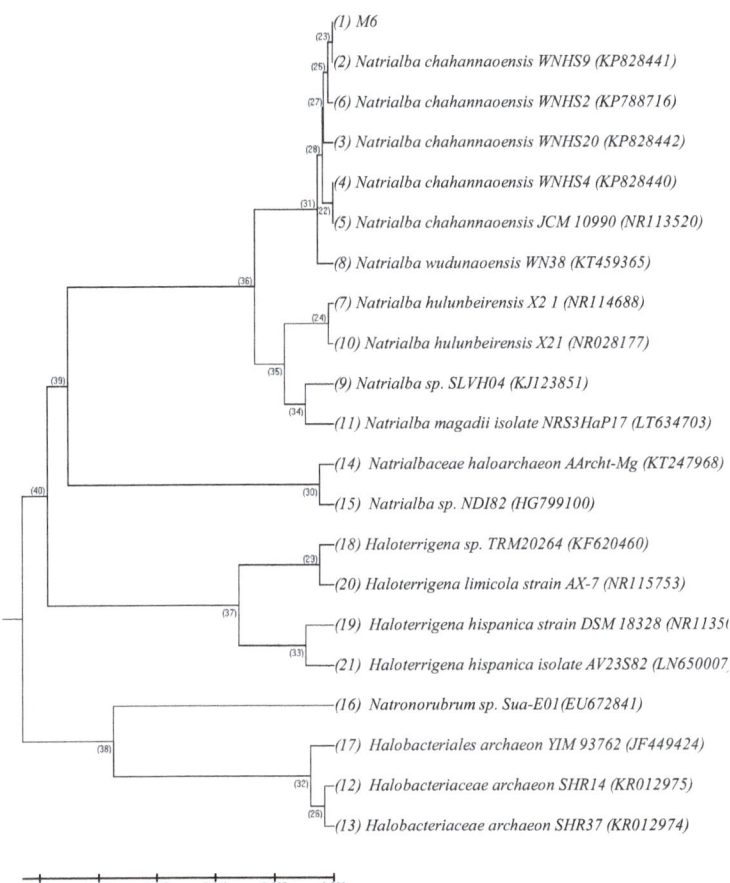

Fig. 5: Phylogenetic tree, based on the 16S rRNA gene sequence comparison, showing the position of the haloalkaliphilic archaeal isolate coded M6 from El Hamra Lake, Wadi El Natrun and the closest relatives. The tree was obtained by the neighbor-joining method. The accession numbers are included in brackets. Bootstrap values are shown at each node of the tree. The scale bar indicates the number of nucleotides substitutions per site.

Growth curve of *Natrialba* sp. M6

Data presented in **Fig. (6)** show the growth of M6 isolate on M5 medium. A lag phase of 3 days was observed before the log phase began, which extended for 10 days then stationary phase started. Afterwards, the organism started to enter the death phase, but the observed increase in OD after this point might be not a true growth (might be artifact caused by cell lysis and the release of pigment). This is obvious from the second curve, which shows that the weight started to decrease gradually after 10 days.

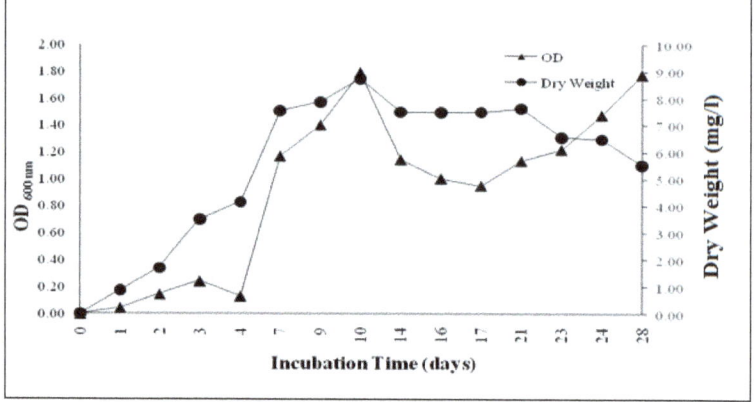

Fig. 6: Growth curve of *Natrialba* sp. M6 based on optical density and dry weight measurements. Cells were grown on M5, pH 11 and incubated at 37°C at 200 rpm for 7 days.

Effect of NaCl concentration, pH and temperature on the growth of *Natrialba* sp. M6

In this experiment, the growth of the bacteria was evaluated with respect to variation in NaCl, pH and temperature. As depicted in **Fig. (7)**, the bacterium failed to grow in the absence or the presence of 10% NaCl (1.71M). Maximum growth (6.9g) was achieved in the presence of 15% NaCl (2.57M) and declined thereafter.

Medium adjusted to pH 7 and 8 supported low growth, which increased gradually with pH increase reaching the maximum value (6 g) at pH 10. Higher pH values (11 and 12) caused a remarkable decrease in growth **(Fig. 8)**. The bacteria appeared to favor growth in a range of temperature (37-50°C) recording the best record (7.5 g) at 45°C **(Fig. 9)**.

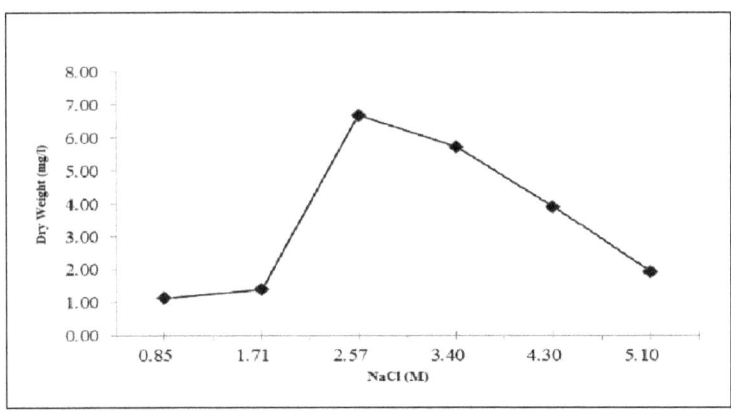

Fig. 7: Growth of *Natrialba* sp. M6 in M5 medium amended with different NaCl concentrations, adjusted to pH 11.0 and incubated shacked at 200 rpm at 37°C for 7 days.

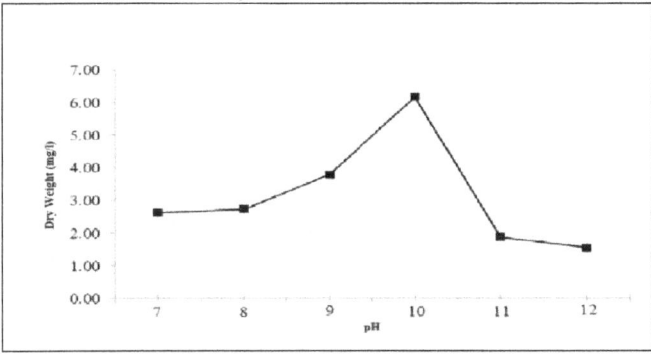

Fig. 8: Growth of *Natrialba* sp. M6 in response to variation in initial pH of the medium. Cells were grown on M5 supplemented with 3.4 M NaCl and incubated shacked at 200 rpm at 37°C for 7 days.

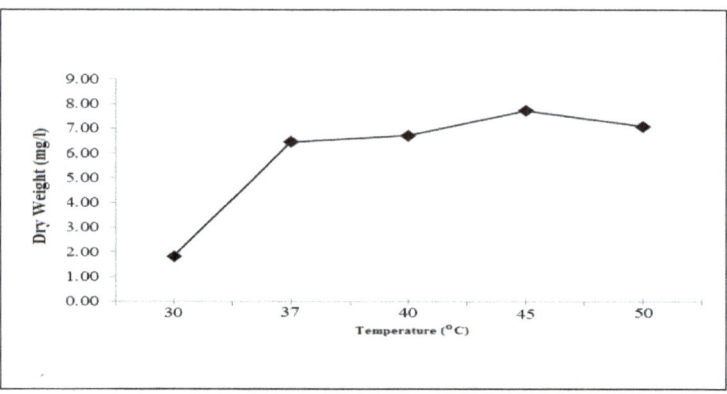

Fig. 9: Effect of incubation temperature on the growth of *Natrialba* sp. M6. Cells were grown on M5 medium, pH 11 supplemented with 3.4 M NaCl, and incubated shacked at 200 rpm for 7 days.

Discussion

Hypersaline habitats occur worldwide and are typical of extreme environments including saline lakes, salterns, and saline and hypersaline soils **(Oren, 2002b, Oren, 2002a)**. Lakes of Wadi El-Natrun, Egypt are considered as one of the most important soda lakes in the world. Halophilic archaea or haloarchaea dominate hypersaline environments. To survive under such extreme conditions, haloarchaea are adapted to function optimally in environments with high salt concentrations and, sometimes, with extreme pH and temperatures. These features make haloarchaea attractive sources of a wide variety of biotechnological products, such as hydrolytic enzymes, with numerous potential applications in biotechnology **(Amoozegar et al., 2017, Torregrosa-Crespo et al., 2018)**.

In the present study, El-Hamra Lake in Wadi El-Natrun was selected for the isolation of halophilic archaeal strains. Isolation of haloarchaea from Wadi El-Natrun was the interest of other researchers **(Hassan et al., 2012)**. 52 isolates were purified, out of which, 37 were proven to be archaeal isolates, based on the PCR product obtained after using the archaeal gene-specific primers **(Valenzuela-Encinas et al., 2008)**. The obtained colonies were pigmented with different shades of red, pink and orange. Pigment production is a typical and characteristic feature of halophilic archaea. In addition, it was observed that a combination of different hydrolytic activities was detected in many strains. Similarly, **(Quadri et al., 2016)** reported that strains belonging to *Nartialba aegyptia and Haloferax mediterranei* exhibited hydrolytic activities. It is widely known that archaea represent an important source of enzymes, including proteases, for applied research as well as for basic enzymology **(De Castro et al., 2006, De Castro et al., 2008, Menasria et al., 2018)**. Protease activities of *Haloferax* and *Natrialba* species had been investigated in previous studies **(Pérez-Pomares et al., 2003, Enache & Kamekura, 2010)**. Moreover, biochemical characterization of all isolates showed lower hydrolytic reactions towards gelatin, Tween 80, casein hydrolysis, as well as indole production.

In the current work, biosurfactant production by the isolated archaea was tested qualitatively using different techniques including surface tension measurement, blood haemolysis, oil-spreading technique and emulsification index as described earlier **(Nitschke et al., 2011)**. Two isolates (M6 and A5) were able to produce biosurfactant as indicated by the emulsification index, and the reduction in the surface tension of cell-free supernatant techniques.

One isolate M6 showing pigment and biosurfactant production was phenotypically characterized via morphological and molecular identification according to **(Anwar & Chauhan, 2011)**. In addition, molecular characterization using BLAST comparisons for the partial sequence of 16S rRNA gene (1322bps) revealed that M6 belongs to the genus *Natrialba*, with 99% similarity to *Natrialba* sp. WNHS9. The 16S-rRNA gene sequence was deposited in the GenBank under accession number (MK063890). *Natrialba aegyptia* was originally isolated from

27

hypersaline soil in Egypt **(Cui et al., 2011)**. In addition, two strains affiliating to *Natrialba taiwanesis* were described in Ouargla, Algeria, but a few studies had reported the presence of this genus in Algerian Sahara, in particularly at Ain Salah **(Xia et al., 2011)**. **(Gana et al., 2011)** reported that ten strains, closely related to *Natrialba aegyptia*, with at least 99% 16S-rRNA gene similarity, were isolated from the hypersaline environments of the Algerian Sahara. In addition, **(Ukani et al., 2011)** isolated and characterized several haloarchaeal strains from Sambhar Lake, Rajasthan, in which saline and alkaline water exist **(Ukani et al., 2011)**.

In the current study, in order to maximize the growth of *Natrialba* sp. M6, pre-optimization process was applied. As a first step, one-variable-at-a-time approach was adopted to determine the main factors affecting the growth of the organism. The strain exhibited optimum growth between 15-20% (w/v) NaCl concentration, and no growth was observed in the absence of NaCl. In a previous study, **(Khemili-Talbi et al., 2015)** reported that *Natrialba* sp. C21 was able to grow in media containing different concentrations of NaCl that ranged between 15–35 % (w/v), and similarly no growth was observed in the absence of NaCl **(Khemili-Talbi et al., 2015)**. Extreme halophiles can grow optimally on a medium supplemented with 3.4–5.1M (20–30%) NaCl **(DasSarma & DasSarma, 2015)**. Their conclusion supports the results of the current study, which revealed the extreme halophilic nature of M6 strain. This demonstrated by the ability of this isolate to grow optimally at 15-20% (w/v) NaCl with a significant growth in the presence of 30% NaCl (w/v). In addition, the organism showed the best growth at pH equal 10, and then the growth was decrease up to pH 11. The favorable temperature for M6 growth was 45°C, these results indicate that the organism is alkaliphilic and moderate thermophilic. In a previous study, **(Khemili-Talbi et al., 2015)** reported that the optimal growth conditions of *Natrialba* sp. C21 were 40°C and pH 7–8.

It was found that cell dry weight of *Natrialba* isolate M6 was approximately 8 mg/L growing on basal medium at 37°C, 200 rpm, pH11.0 and 20% NaCl. Meanwhile, the cell dry weight reached it maximum value (approximately 10.5 mg/L) when the isolate was grown on the pre-optimized medium under the following conditions 45°C, pH 10.0, rpm 200, and 15% NaCl. **(Manikandan et al., 2009)** reported that the growth and biomass production of *Halobacterium salinarum* VKMM 013 that was carried out using response surface method reached of 0.746 g/L dry weight was in agreement with the predicted cell concentration using the following conditions: 35 g/L of KCl, 9.70 g/Lof MgSO$_4$, 13.38 g/Lof gelatin and 12.00 g/L soluble starch in nutrient broth supplemented with artificial seawater and 20% (w/v) of NaCl.

Archaeal strains isolated by **(Quadri et al., 2016)** including *Natrialba* required at least 12% NaCl concentration, and grew optimally at the following conditions; 25% of NaCl, 40°C and pH 7.5–8. The difference in the optimum pH refers to the alkaliphilism nature of the M6 strain.

The search for biosurfactants in extremophiles is a promising research area, since the biosurfactants are important for the adaptation of these organisms and increase their stability in the adverse environments. These characteristics make these biosurfactants are suitable for several biotechnological applications **(Kebbouche-Gana et al., 2013)**. There are very few reports on biosurfactant production in hypersaline environments. Other studies reported the isolation of biosurfactant-producing haloalkaliphilic archaea from Wadi- El-Natrun, Egypt **(Hegazy et al., 2022)**. Other biosurfactant was also produced from *Haloferax* sp. MSNC14 **(Djeridi et al., 2013)**. **(Kebbouche-Gana et al., 2013)** reported the production of biosurfactant from *Natrialba* sp. strain E21, an extremely halophilic bacterium isolated from a solar saltern (Ain Salah, Algeria). **(Khemili-Talbi et al., 2015)** reported that *Natrialba* sp. C21 was able to produce biosurfactant to facilitate the uptake of aromatic hydrocarbons and their biodegradation even at high salt concentration.

Conclusion

Different archaeal isolates have been isolated from El-Hamra lake in Wadi El-Natrun. Depending on the archaeal special primers, the obtained archaeal isolates were differentiated from the non-archaeal isolates. According to the chemical analysis of the lake and its slat content, the obtained isolates were classified as halophilic archaea. The ability of the selected isolate (that has been identified as *Natrialba* sp.) to produce hydrolytic enzymes and biosurfactants, in addition to its physiological growth conditions (NaCl concentration, pH, and temperature) was investigated.

References

Abd-El-Malek Y, Rizk S, 1963. Bacterial sulphate reduction and the development of alkalinity. III. Experiments under natural conditions in the Wadi Natrun. *Journal of Applied microbiology* **26**, 20-6.

Ai S, V S, J W, *et al.*, 2003 TM4: a free, open-source system for microarray data management and analysis. *Biotechniques* **34**, 374-8.

Amoozegar MA, Siroosi M, Atashgahi S, Smidt H, Ventosa A, 2017. Systematics of haloarchaea and biotechnological potential of their hydrolytic enzymes. *Microbiology* **163**, 623-45.

Anwar T, Chauhan R, 2011. *Mining Bacterial Strains from a Rock Salt Mine for Halotolerance Genes and Enzymes*: Jaypee University of Information Technology, Solan, HP.

Cheesbrough M, 1981. *Medical laboratory manual for tropical countries*. M. Cheesbrough, 14 Bevills Close, Doddington, Cambridgeshire, PE15 OTT.

Clarke SK, 1953. A simplified plate method for detecting gelatine-liquefying bacteria. *Journal of Clinical Pathology* **6**, 246.

Collee J, Miles R, Watt B, 1996. Collee JG, Fraser AG, Marmion BP, Simmons A. *Mackie and McCartney Practical Medical Microbiology. 14th. Churchill Livingstone.*

Cui H-L, Yang X, Mou Y-Z, 2011. Salinarchaeum laminariae gen. nov., sp. nov.: a new member of the family Halobacteriaceae isolated from salted brown alga Laminaria. *Extremophiles* **15**, 625-31.

Dassarma S, Dassarma P, 2015. Halophiles and their enzymes: negativity put to good use. *Current opinion in microbiology* **25**, 120-6.

De Castro RE, Maupin-Furlow JA, Giménez MI, Herrera Seitz MK, Sánchez JJ, 2006. Haloarchaeal proteases and proteolytic systems. *FEMS microbiology reviews* **30**, 17-35.

De Castro RE, Ruiz DM, Giménez MI, Silveyra MX, Paggi RA, Maupin-Furlow JA, 2008. Gene cloning and heterologous synthesis of a haloalkaliphilic extracellular protease of Natrialba magadii (Nep). *Extremophiles* **12**, 677-87.

Djeridi I, Militon C, Grossi V, Cuny P, 2013. Evidence for surfactant production by the haloarchaeon Haloferax sp. MSNC14 in hydrocarbon-containing media. *Extremophiles* **17**, 669-75.

Enache M, Kamekura M, 2010. Hydrolytic enzymes of halophilic microorganisms and their economic values. *Rom J Biochem* **47**, 46-59.

Falchetti A, Di Stefano M, Marini F*, et al.*, 2005. Segregation of a M404V mutation of the p62/sequestosome 1 (p62/SQSTM1) gene with polyostotic Paget's disease of bone in an Italian family. *Arthritis research & therapy* **7**, 1-7.

Gana ML, Kebbouche-Gana S, Touzi A*, et al.*, 2011. Antagonistic activity of Bacillus sp. obtained from an Algerian oilfield and chemical biocide THPS against sulfate-reducing bacteria consortium inducing corrosion in the oil industry. *Journal of Industrial Microbiology and Biotechnology* **38**, 391-404.

Hall TA. BioEdit: a user-friendly biological sequence alignment editor and analysis program for Windows 95/98/NT. *Proceedings of the Nucleic acids symposium series, 1999*: Oxford, 95-8.

Hassan HA, Rizk NM, Hefnawy M, Awad AM, 2012. Isolation and characterization of halophilic aromatic and chloroaromatic degrader from Wadi El-Natrun Soda lakes. *Life Sci J* **9**, 1565-70.

Hegazy GE, Abu-Serie MM, Abou-Elela G, *et al.*, 2022. Bioprocess development for biosurfactant production by Natrialba sp. M6 with effective direct virucidal and anti-replicative potential against HCV and HSV. *Scientific reports* **12**, 16577.

Holmes DS, Quigley M, 1981. A rapid boiling method for the preparation of bacterial plasmids. *Analytical Biochemistry* **114**, 193-7.

Kebbouche-Gana S, Gana ML, Ferrioune I, *et al.*, 2013. Production of biosurfactant on crude date syrup under saline conditions by entrapped cells of Natrialba sp. strain E21, an extremely halophilic bacterium isolated from a solar saltern (Ain Salah, Algeria). *Extremophiles* **17**, 981-93.

Khemili-Talbi S, Kebbouche-Gana S, Akmoussi-Toumi S, Angar Y, Gana ML, 2015. Isolation of an extremely halophilic arhaeon Natrialba sp. C21 able to degrade aromatic compounds and to produce stable biosurfactant at high salinity. *Extremophiles* **19**, 1109-20.

Manikandan M, Pašić L, Kannan V, 2009. Optimization of growth media for obtaining high-cell density cultures of halophilic archaea (family Halobacteriaceae) by response surface methodology. *Bioresource technology* **100**, 3107-12.

Menasria T, Aguilera M, Hocine H, *et al.*, 2018. Diversity and bioprospecting of extremely halophilic archaea isolated from Algerian arid and semi-arid wetland ecosystems for halophilic-active hydrolytic enzymes. *Microbiological Research* **207**, 289-98.

Mesbah NM, Abou-El-Ela SH, Wiegel J, 2007. Novel and unexpected prokaryotic diversity in water and sediments of the alkaline, hypersaline lakes of the Wadi An Natrun, Egypt. *Microbial ecology* **54**, 598-617.

Moore E, Arnscheidt A, Krüger A, Strömpl C, Mau M, 2004. Section 1 update: Simplified protocols for the preparation of genomic DNA from bacterial cultures. In: Kowalchuk GA, De Bruijn FJ, Head IM, Akkermans AD, Van Elsas JD, eds. *Molecular Microbial Ecology Manual.* Springer Netherlands, 1905-19.

Nitschke M, Costa SG, Contiero J, 2011. Rhamnolipids and PHAs: Recent reports on Pseudomonas-derived molecules of increasing industrial interest. *Process Biochemistry* **46**, 621-30.

Oren A, 2002a. Diversity of halophilic microorganisms: environments, phylogeny, physiology, and applications. *Journal of Industrial Microbiology and Biotechnology* **28**, 56-63.

Oren A, 2002b. Molecular ecology of extremely halophilic Archaea and Bacteria. *FEMS microbiology ecology* **39**, 1-7.

Pérez-Pomares F, Bautista V, Ferrer J, Pire C, Marhuenda-Egea F, Bonete M, 2003. α-Amylase activity from the halophilic archaeon Haloferax mediterranei. *Extremophiles* **7**, 299-306.

Poli A, Finore I, Romano I, Gioiello A, Lama L, Nicolaus B, 2017. Microbial diversity in extreme marine habitats and their biomolecules. *Microorganisms* **5**, 25.

Quadri I, Hassani II, L'haridon S, Chalopin M, Hacène H, Jebbar M, 2016. Characterization and antimicrobial potential of extremely halophilic archaea isolated from hypersaline environments of the Algerian Sahara. *Microbiological Research* **186**, 119-31.

Salgaonkar BB, Mani K, Bragança JM, 2013. Accumulation of polyhydroxyalkanoates by halophilic archaea isolated from traditional solar salterns of India. *Extremophiles* **17**, 787-95.

Sanbrook J, Fritsch E, Maniatis T, 1989. Molecular cloning: a laboratory manual. *Cold Spring Harbor, NY, Cold Spring Harbor Laboratory* **11**, 31.

Sanger F, Nicklen S, Coulson AR, 1977. DNA sequencing with chain-terminating inhibitors. *Proceedings of the national academy of sciences* **74**, 5463-7.

Shakya M, Quince C, Campbell JH, Yang ZK, Schadt CW, Podar M, 2013. Comparative metagenomic and rRNA microbial diversity characterization using archaeal and bacterial synthetic communities. *Environmental microbiology* **15**, 1882-99.

Simmons JS, 1926. A culture medium for differentiating organisms of typhoid-colon aerogenes groups and for isolation of certain fungi. *The Journal of Infectious Diseases*, 209-14.

Tamura K, Dudley J, Nei M, Kumar S, 2007. MEGA4: molecular evolutionary genetics analysis (MEGA) software version 4.0. *Molecular biology and evolution* **24**, 1596-9.

Tindall B, Ross H, Grant W, 1984. Natronobacterium gen. nov. and Natronococcus gen. nov., two new genera of haloalkaliphilic archaebacteria. *Systematic and Applied Microbiology* **5**, 41-57.

Torregrosa-Crespo J, Montero Z, Fuentes JL, *et al.*, 2018. Exploring the valuable carotenoids for the large-scale production by marine microorganisms. *Marine drugs* **16**, 203.

Ukani H, Purohit MK, Georrge JJ, Paul S, Singh SP, 2011. HaloBase: development of database system for halophilic bacteria and archaea with respect to proteomics, genomics & other molecular traits.

Valenzuela-Encinas C, Neria-González I, Alcántara-Hernández RJ, *et al.*, 2008. Phylogenetic analysis of the archaeal community in an alkaline-saline soil of the former lake Texcoco (Mexico). *Extremophiles* **12**, 247-54.

Vijayanand S, Hemapriya J, Selvin J, Kiran S, 2010. Production and optimization of haloalkaliphilic protease by an extremophile-Halobacterium sp. Js1, isolated from thalassohaline environment. *Global J Biotechnol Biochem* **5**, 44-9.

Widdel F, 2007. Theory and measurement of bacterial growth. *Di dalam Grundpraktikum Mikrobiologie* **4**, 1-11.

Xia W-J, Dong H-P, Yu L, Yu D-F, 2011. Comparative study of biosurfactant produced by microorganisms isolated from formation water of petroleum reservoir. *Colloids and Surfaces A: Physicochemical and Engineering Aspects* **392**, 124-30.

Optimization of Biosurfactant and Pigment Production by *Natrialba* sp. M6: Characterization, Biological Activities, and Potential Applications

Abstract

This study aimed to investigate the optimum formula for maximum pigment production by *Natrialba* sp. M6 and analyze the properties and biological activities of the produced biosurfactant and pigment. Fourier-Transform Infrared Spectroscopy (FT-IR) analysis showed the presence of specific chemical groups in the pigment extract, indicating the presence of aliphatic compounds. Raman spectroscopy analysis confirmed the presence of orange carotenoids in the pigment extract. The cytotoxicity and selectivity index (SI) values indicated higher selectivity of the pigment extract towards cancer cells compared to chemotherapy and biosurfactant. Flow cytometry analysis and fluorescence microscopy confirmed the induction of apoptosis in cancer cells by the pigment extract. In vitro inhibition of albumin denaturation assay revealed anti-inflammatory activity of the pigment extract at high concentrations. The pigment extract exhibited higher antioxidant capacity compared to the biosurfactant extract. Moreover, both the pigment and biosurfactant extracts showed anti-biofouling activity by reducing bacterial biofilms. Overall, these results demonstrate the potential of *Natrialba* sp. M6 biosurfactant and pigment extract for various applications including anticancer, anti-inflammatory, antioxidant, and anti-biofouling activities.

Keywords: Biosurfactant, Pigment, Anticancer, Anti-inflammatory, Anti-biofouling

Introduction

Biosurfactants and pigments produced by microorganisms have garnered significant attention due to their diverse applications in various industries, including bioremediation, pharmaceuticals, food, and cosmetics. *Natrialba* sp. M6, a halophilic archaeon, has emerged as a promising candidate for the production of biosurfactants and pigments. This study aims to optimize the production of biosurfactants and pigments by *Natrialba* sp. M6, characterize their properties, evaluate their biological activities, and explore their potential applications. *Natrialba* sp. M6 belongs to the genus *Natrialba*, which is known for its ability to thrive in highly saline environments. Halophilic microorganisms have adapted to survive in extreme conditions, including high salt concentrations, making them a valuable resource for industrial applications. Moreover, the ability of *Natrialba* sp. M6 to produce biosurfactants and pigments adds to its potential as a bioresource for various industries **(Hegazy et al., 2020)**. Biosurfactants are amphiphilic compounds produced by microorganisms that exhibit surface-active properties. They have unique advantages over chemically synthesized surfactants, such as their biodegradability, low toxicity, and environmental compatibility. Biosurfactants play a vital role in various applications, including oil recovery, enhanced oil spill remediation, and emulsification of hydrophobic compounds. Additionally, they possess antimicrobial, anti-adhesive, and anti-biofilm properties, making them attractive for applications in healthcare and biotechnology **(Hegazy et al., 2022)**. Pigments produced by microorganisms have gained immense interest due to their diverse colors, stability, and potential use as natural colorants. Microbial pigments offer advantages over synthetic pigments, such as their biodegradability, low toxicity, and ease of production. These pigments find applications in the food, cosmetic, and textile industries **(Hegazy et al., 2020)**. Furthermore, recent studies have highlighted their antioxidant, antimicrobial, and anticancer properties, expanding their potential uses in healthcare and pharmaceuticals. To optimize the production of biosurfactants and pigments by *Natrialba* sp. M6, various parameters need to be considered, including culture conditions, carbon and nitrogen sources, pH, temperature, and incubation time. Optimization studies can enhance the yield and quality of biosurfactants and pigments, making them more economically viable for industrial applications. Characterization of the biosurfactants and pigments produced by *Natrialba* sp. M6 is essential to understand their chemical composition, molecular structure, thermal stability, and surface-active properties. Analytical techniques such as mass spectrometry, nuclear magnetic resonance (NMR), and Fourier-transform infrared spectroscopy (FTIR) can provide valuable insights into the chemical nature and functional groups present in these compounds. Characterization studies will enable researchers to determine the suitability of *Natrialba* sp. M6-derived biosurfactants and pigments for specific applications. Furthermore, evaluating the biological activities of biosurfactants and pigments produced by *Natrialba* sp. M6 is crucial to assess their potential applications in various fields. Antimicrobial, anti-inflammatory, antioxidant, and anticancer activities are among the key biological properties that can be investigated. Understanding the biological activities of these compounds will aid in identifying their potential as therapeutic agents, functional food additives, or cosmetic ingredients. In conclusion, this research aims to optimize the production of biosurfactants and pigments by *Natrialba* sp. M6, characterize their properties, evaluate their biological activities, and explore their potential applications. The findings from this research will contribute to the development of sustainable and eco-friendly alternatives for surfactants and pigments in various industries. Moreover, *Natrialba* sp. M6-derived biosurfactants and pigments hold promise for their potential applications in bioremediation, pharmaceuticals, food, and cosmetics, offering novel and sustainable solutions for various industrial challenges **(Hegazy et al., 2022).**

Methods

Screening for the production of biosurfactants

A seed culture was prepared for each bacterium until OD ~ 0.9. A standard inoculum of 1 ml was used to inoculate 250 ml flasks, each containing 100 ml basal medium (M5) adjusted to pH 11. Flasks were incubated shake, 200 rpm at 37°C for 7days incubation.

Surface tension measurement

Surface tension was measured as described by **(Haba et al., 2000)**. Bacterial cultures were centrifuged at 14000 rpm for 10 min and the surface tension of each supernatant was measured using a tensiometer (TDI, Lauda, Germany) and expressed as mN/m using water as a reference.

Emulsification index (% EI$_{24}$)

Emulsification activity was measured using the method described by **(Iqbal et al., 1995)**. Four ml of gas oil were added to

4 ml of the culture broth in a graduated screw cap test tube, and vortexed at high speed for 2 min. The emulsion stability was determined after 24 h, and the emulsification index (% EI_{24}) was calculated by dividing the measured height of the emulsion layer by total height of the mixture and multiplying by 100.

(Height of emulsified layer/total height of the layers) *100

Oil spreading technique

Fifteen µl of crude oil were placed on the surface of distilled water (40 µl) in a petri dish (150 mm in diameter). Then, 10 µl of the culture supernatant were gently put on the centre of the oil film. Appearance of clear area under visible light were scored as positive, while absence of this clear area were scored as negative **(Morikawa et al., 2000)**.

Haemolytic activity

Archaeal strains were screened on blood agar plates containing 5% (v/v) sheep blood and incubated at 37°C for 24 h. Haemolytic activity was detected as the presence of a definite clear zone around colonies which is indicative of surfactant biosynthesis **(Youssef et al., 2004)**.

Production of biosurfactant and pigment by *Natrialba* sp. M6

In a preliminary experiment, *Natrialba* sp. M6 was allowed to grow in the basal medium (pH 11, temperature 37° C and 200 g/l NaCl (3.4M)) and pre-optimized medium (pH 10, temperature 45°C, and NaCl 150 g/l (2.57M)). Both biosurfactant and pigment production were monitored and detected in association to growth (OD_{600nm}).

Optimization using experimental factorial designs

The optimization process was carried out in three sequential steps based on statistically designed experiments. First: elucidation of medium components using a two-level screening design. Second: optimization of the most significant components by Box-Behnken statistical experimental design, creating a mathematical model expressing the relationship between optimized factors. Finally, verification of the model and monitoring the production pattern were carried out **(Abdel-Fattah et al., 2007, Abdel-Fattah et al., 2009)**.

Plackett-Burman Design (Plackett & Burman, 1946)

This experiment was applied to biosurfactant activity. Plackett-Burman experimental design was used to evaluate the relative significance of 14 culture factors, including medium components and other culture parameters that affect growth and biosurfactant yield of *Natrialba* sp. M6, respectively. Based on the Plackett-Burman factorial design, each factor was examined at 2 levels: '-1' for the low level, and '+1' for the high level. In addition, the matrix designs of the tested factors were screened in 16 experimental trials. All trials were performed in 250 mL flasks containing 100 ml of the medium. The response was detected through the measurement of the surface tension in mN/m using a tensiometer.

Plackett-Burman design is based on a first order model:

$$Y = \beta_0 + \sum \beta_i x_i$$

Where Y is the response (reciprocal of corrected surface tension), β_0 is the model intercept, β_i is the variable estimate and x_i represent the variable. The Pareto plot best demonstrates the results of Plackett-Burman design since it illustrates the absolute relative significance of variables independent on their nature **(Plackett & Burman, 1946).**

Verification experiment

A pre-optimization step should be done for subsequent optimization step. In this step, a pre-optimization formula was prepared, where the most significant variables were fixed at their optimum levels obtained from Plackett-Burman design. On the other hand, the other variables with a negative effect value were fixed at their '-1' coded values, and those with a positive effect value were fixed at their '+1' coded values. The purpose of this step is to confirm the results of Plackett-Burman design and to construct the basic formula for further optimization step.

- ### Response surface methodology (Box & Behnken, 1960; Abdel-Fattah *et al.*, 2007b)

After estimating the relative significance of independent variables, the most significant variables were selected for further determination of their optimal level with respect to reciprocal of corrected surface tension as a response expressing the biosurfactant yield. For this reason, Box-Behnken Design, which is a response surface methodology, was applied. This optimization process involves three main steps: performing the statistically designed experiments, estimating the coefficients in a mathematical model and predicting the response and checking the adequacy of the model. The four significant variables elucidated through Plackett-Burman experimental design for *Natrialba* sp. M6 isolate were pH (X_1), glycerol (X_2), agitation (X_3) and NaCl (X_4), while those for pigment were NaCl (X_1), pH (X_2) and culture age (X_3). The low, middle and high levels of each variable were designated as -1, 0 and +1, respectively. A design matrix for 27 trials for biosurfactant and 15 trials for pigment, along with the natural values for the factors were constructed. The mean values of reciprocal of corrected surface tension were calculated. To predict the optimal point, a second order polynomial function was fitted to correlate the relationship between the independent variables and the response. The equation for the four factors for biosurfactant was as follows:

$$Y = \beta_0 + \beta_1(X_1) + \beta_2(X_2) + \beta_3(X_3) + \beta_4(X_4) + \beta_{12}(X_1X_2) + \beta_{13}(X_1X_3) + \beta_{14}(X_1X_4) \quad \beta_{23}(X_2X_3) + \beta_{24}(X_2X_4) + \beta_{34}(X_3X_4) + \beta_{11}(X_1)^2 + \beta_{22}(X_2)^2 + \beta_{33}(X_3)^2 + \beta_{44}(X_4)^2$$

The equation for the three factors for pigment was as follows:

$$Y = \beta_0 + \beta_1(X_1) + \beta_2(X_2) + \beta_3(X_3) + \beta_{12}(X_1X_2) + \beta_{13}(X_1X_3) + \beta_{23}(X_2X_3) + \beta_{11}(X_1)^2 + \beta_{22}(X_2)^2 + \beta_{33}(X_3)^2$$

where, Y is the predicted response (biosurfactant and pigment), β_0 is constant, β_1, β_2, β_3 and β_4 are linear coefficients, β_{12}, β_{13} and β_{23} are cross product coefficients, and β_{11}, β_{22}, β_{33} and β_{44} are quadratic coefficients. Variables maximal predicted response and coefficients calculations were carried out using Microsoft Excel 2007.

Statistical analysis of the data

The data was subjected to multiple linear regressions using Microsoft Excel to estimate *t*-value, *p*-value and confidence level. The significance level (*p*-value) is determined using the *t* test. The *t*-test for any individual effect allows an evaluation of the probability of finding the observed effect purely by chance. If this probability is sufficiently small, the idea that the effect was caused by varying the level of the variable under test is accepted. Confidence level is an expression of the *p*-value is percent. Optimal value of activity and specific activity were estimated using the *solver* function of Microsoft Excel tools. The simultaneous effects of the three most significant independent factors on each response were visualized using three-dimensional graphs generated by Statistica 5.0 software.

Model verification

The optimal conditions obtained from the optimization experiments were verified experimentally and were then compared to the data calculated from the model.

Recovery and characterization of biosurfactant

Natrialba sp. M6 was harvested by centrifugation (16,000g, 20min) from the culture broth. For biosurfactant extraction, the cell-free supernatant was acidified to pH 2 using 6N HCl solution and then kept at 4°C overnight. The precipitate was collected by centrifugation (22,000 g, 20 min). The residue of biosurfactant after precipitation was weighed and dissolved in known volume of 0.1 M sodium bicarbonate **(Thaniyavarn et al., 2006).**

Characterization of biosurfactant

- ### Gas chromatography-mass spectrometry (GC-MS) analysis

GC-MS analysis was performed according to **(Jerković et al., 2015)**. Using an Agilent technologies (GC) equipped with mass selective detector (MS), HP-5MS at the marine pollution lab, National Institute of Oceanography and Fisheries, Alexandria, Egypt, 5% phenyl methyl siloxane capillary column of dimensions 30.0m×250µm×0.25µm was used using helium as carrier gas at 1 ml/ min. The column temperature was programmed initially at 90°C for 1 min, followed by an increase of 8°C/min to 205°C, then 5°C/min to 240°C, then 8°C/min to 300°C. The MS was operating at 70 eV. The constituents were identified by comparison of their mass spectral data with those standard compounds from NIST (National Institute of Standards and Technology) Spectral Library.

- **Protein measurement using Folin reagent (Oliver *et al.*, 1951)**

In test tubes, 2.5 ml of alkaline copper solution were added to 500 µl of biosurfactant, standards and blank, mixed well and allowed to stand for 10 min or longer at room temperature. Then, 250 µl of diluted Folin reagent was added rapidly and mixed within a second or two. After 20 min, the samples were read in a colorimeter or spectrophotometer (double beam metertech UV/Vis spectrophotomer SP-8001) at 750 nm and the protein concentration was calculated from a standard curve. This was done at the marine chemistry lab, National Institute of Oceanography and Fisheries, Alexandria, Egypt. Standard curve was prepared using human serum with concentrations from 100 to 500 µg/ml. These, in turn may be checked against a standard solution of crystalline bovine albumin.

- **Determination of total lipids (Christopher *et al.*, 1971)**

In test tubes containing 500 µl of biosurfactant, standards and blank, 250 µl of concentrated sulfuric acid were added and mixed well. Test tubes were placed in boiling water for about 5 min, then 5 ml of the phospho-vanillin reagent were added to each tube, mixed well, and incubated at 37°C in water bath for 15 min. The tubes were cooled for about 5 min and then within 30 min the absorbance was measured at 540 nm using a double beam meter UV/Vis spectrophotomer SP-8001. Standards were prepared from cholesterol standard with concentrations from 50 to 200 mg/dl at the marine chemistry lab, National Institute of Oceanography and Fisheries, Alexandria, Egypt.

- **Determination of carbohydrates (Michel *et al.*, 1955)**

To 600 µl of biosurfactant sample, standards and blank, a volume of 600 µl of phenol (5%w/v) were added and mixed well with 3 ml concentrated sulfuric acid. The test tubes were left at room temperature for 30 min then measured at 490 nm using a double beam meter tech UV/Vis spectrophotomer (SP-8001). Standards series were prepared from D-glucose with concentrations 20-100 mg/l at the Marine Chemistry Lab, National Institute of Oceanography and Fisheries, Alexandria, Egypt.

Pigment extraction and characterization (Davood *et al.*, 2014)

50 ml of the culture broth were centrifuged at 10.000 g for 20 min at 4°C. The supernatant was discarded and 50 ml of distilled water added to the pellets and then kept at 4°C overnight. A mixture of acetone-methanol (7:3 v/v) containing butylhydroxytoluene (BHT) (0.1% as antioxidant) was added to the pellets. Successive extractions were carried out until both solvents and cells were colourless and then again were centrifuged. The solvent was evaporated at 45°C overnight and the pigment were dissolved in 50 ml acetone containing 0.1% BHT. The colored extract was analyzed by scanning the absorbance in the wavelength region of 200-700 nm using the Perkin Elimer UV/VIS Lambda EZ 201, USA Spectrophotometer, where λmax was found at 350 nm.

Absorption –pigment weight correlation

The correlation curve between Absorption λmax (350 nm) and pigment concentrations (µg/ml) was plotted. The correlation factor of absorption (λmax), pigment concentrations ranged from 300-6000 µg/ml was determined from the linear relation **(Fig. 1)**. Subsequently, the correlation equation was applied in the optimization experiment to convert the absorbance into concentration in RSM optimization experiment **(Thrane et al., 2015)**.

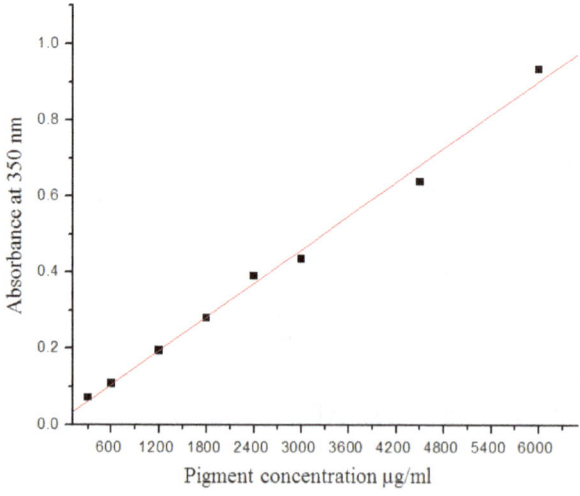

Linear Fit on linearized scales.
Y scale (Y) = A + B * x scale (X)
where scale (x) is the current axis scale function.

Parameter	Value	Error
A	0.01806	0.01497
B	1.47148	4.85223

R	SD	N
0.99675	0.025288	<0.0001

Fig. 1: The correlation between absorption and pigment weight.

Characterization of pigment (Kirti *et al.*, 2014)
- ### Fourier-transform infrared spectroscopy (FTIR) analysis

The molecular structure of the pigment was partially identified using Peak Find-Memory-27 spectrophotometer, at the central lab, City of Scientific Researches and Technological Applications, Borg El-Arab, Alexandria, Egypt. A mixture of about 1 mg of the tested material and 300 mg of pure dry potassium bromide (KBr) were pressing into discs. The measurements were carried out at infrared spectra between 400-4000 nm.

- ### Raman spectroscopy

A sample (785 nm) with power 50 mW with a wide range 400-4000 cm was examined using Raman Senteraa Model, at the Central Lab, City of Scientific Researches and Technological Applications, Borg El-Arab, Alexandria, Egypt. The laser irradiates the object in the optical microscope (laser spot = 2μm), the scattered light from the sample was collected by the optics of the microscope passing through holographic filters, pinhole, and monochromatic to be detected by a charge-coupled device (CCD).

Applications of *Natrialba* sp. M6 biosurfactant and pigment

- ### Determination of cytotoxicity on human normal cells and cancer cell lines

Normal human lung fibroblast cells (Wi-38) and colon cancer line (Caco-2) were maintained as adherent cell cultures in Dulbecco's modified Eagle medium (DMEM) containing 10 %

45

Fetal Bovine Serum (FBS). Liver cancer cell line (HepG-2), breast cancer cell line (MCF-7) and cervical cancer cell line (Hela) were cultured in Roswell Park Memorial Institute medium (RPMI) 1640 supplemented with 10% FBS and 4 mM L-glutamine (Lonza, USA). Human normal cells and four cancer cell lines were routinely maintained as adherent cell cultures at 37°C in a humidified air incubator containing 5% CO_2. Each cell line was seeded in 25 cm^2 culture flask and incubated in 5% CO_2 incubator. When reach confluent each cell type was sub-cultured using trypsinization into 75 cm^2 culture flask for two weeks before assays. Cell viability was assessed using the dye exclusion method of **(Louis & Siegel, 2011)**. It is based on the principle that live cells possess intact cell membranes that exclude certain dyes, such as trypan blue, whereas dead cells do not. A portion of the cell suspension (e.g., 50 µl) was mixed with an equal volume of 0.5% trypan blue staining solution and loaded onto haemocytometer. Both viable "unstained" and nonviable "stained" cells were counted in each of the four corner quadrants (A, B, C, D).

$$N / ml = \left(\frac{A+B+C+D}{4} \right) \times 10^4 \times D$$

N: Number of viable or nonviable cells
D: Sample dilution (1:1 with the trypan blue).

% of cell viability= $\dfrac{\text{Number of viable cells}}{\text{Total number of cells}} \times 100$

At least 90% of the cells must be viable in order to use the cells for assays.

The sensitivity of normal and tumor cells to extracts was evaluated using MTT assay, as described by **(Mosmann, 1983)**. Cell suspensions (4×10^3 cells/well) were seeded into sterile 96-well flat-bottomed cell culture plates and allowed to attach for 24 h. Then serial concentrations of the extracts and standard chemotherapy (5-fluorouracil) were incubated with four cancer cell lines and plates were incubated at 37°C in 5% CO_2 incubator for 72 h. After incubation in 5% CO_2 incubator, 20 µl of MTT solution (5 mg/ml in PBS, pH 7) were added to each well and incubated at 37°C for 4 h. MTT solution was then removed after centrifugation at 2000 rpm

46

for 10 min and the insoluble blue formazan crystals trapped in cells were solubilized with 150 µl of 100% DMSO at 37°C for 10 min. The absorbance of each well was measured with a microplate reader at 570 nm **(Louis & Siegel, 2011)**.

$$\text{The percentage of normal cell proliferation} = \left(\frac{OD_E - OD_B}{OD_C - OD_B} \right) \times 100$$

The inhibition of growth rate of treated cancer cells was calculated according to the following formula:

$$\text{The inhibition rate (\%)} = 100 - \left(\frac{OD_E - OD_B}{OD_C - OD_B} \right) \times 100$$

Where:

OD_E: The mean of absorbance of extract - treated cancer cells

OD_B: The mean of absorbance of blank control

OD_C: The mean of absorbance of control untreated cancer cells

The half maximal inhibitory concentration (IC_{50}), and the safe dose (EC_{100}) of extracts were estimated by the Graph-pad Instat software. Moreover, the selectivity index (SI) that defined as the ratio of the IC_{50} on normal cells versus cancer cell lines was estimated.

Additionally, cellular morphological changes before and after treatment were investigated using phase contrast inverted microscope (Olympus, Japan).

o **Analysis of extract-treated cancer cell death effect using flow cytometry and fluorescence phase contrast microscope (Louis & Siegel, 2011)**

Different human cancer cell lines including Caco-2, MCF-7, HepG2 and Hela cell lines were treated with more effective safe anticancer extract and 5-fluorouracil (FU) at their IC_{50} dose. After 48 h, cells were harvested by trypsinization, washed with cold PBS and resuspended with 200 µl 1× binding buffer, followed by incubation with 5 µl of Annexin V- biotin (Sigma, USA) and 5 µl of propidium iodide (Sigma, USA) for 15 min in dark. After staining, cells were washed with 1× binding buffer, fixed with 2% formaldehyde in PBS for 15 min and washed once with PBS. Then 5 µg/ml of streptavidine-fluorescein (Sigma, USA) were added to cells for 15 min, and cells were collected by centrifugation and resuspended in PBS. The cell death rates were detected by flow cytometry (Ex = 488 nm; Em = 530 nm) using FITC signal detector (FL1) and the phycoerythrin emission signal detector (FL2) for estimation of the percentage annexin-stained apoptotic cells and propidium iodide (PI)-stained necrotic cells, respectively.

○ **Fluorescence phase contrast microscope detection of cell death**

Anti-tumor activity of more effective extract was assayed by two fluorescents nuclear staining using ethidium bromide and acridine orange dyes **(Louis & Siegel, 2011)**. Different cancer cells were seeded in 96 well plates and allow attaching for 24 h. Then, cancer cells were treated with IC_{50} doses of more effective extract and FU for 72 h in 5% CO_2 incubator. After washing the cells three times with cold PBS, they were stained with 100 μg/ml ethidium bromide and 100 μg/ml acridine orange dyes (prepared freshly), and examined after 15 min using the fluorescence phase contrast microscope (Olympus, Japan).

- **In vitro antiviral activity**
○ **Isolation of peripheral blood mononuclear cells (PBMCs)**

The blood samples were collected from healthy volunteers in heparinized tubes and then diluted with equal volume of RPMI-1640 medium, and carefully layered on ficoll-hypaque, then centrifuged at 2000 rpm, 25°C for 30 min. The undisturbed PBMCs layer (interface) was carefully transferred out, washed with 40 ml RPMI-1640 medium, and centrifuged at 1650 rpm for 10 min. Finally, the supernatant was removed and the cells were suspended in RPMI-1640 medium containing 10% FBS and counted using trypan blue stain **(Mosmann, 1983).**

○ **Determination of cytotoxicity of biosurfactant and pigment on PBMCs**

The cytotoxicity assay was done according to **(Mosmann, 1983).** Mononuclear cells ($1x10^6$ cells/ml) were resuspended gently in RPMI medium containing 10% FBS and the cell count was adjusted to $1x10^5$ cells/ml of the selected culture medium. To each well of the 96 wells in the micro titre plate, $1x10^5$ mononuclear cells were seeded. Then cells were incubated with and without the serial dilutions of extracts and currently used antiviral drug (sovaldi). After 72 h incubation in 5% CO_2 incubator, 20 μl of 5 mg/ml MTT solution (Sigma, USA) were added to each well and the plate was incubated at 37°C for 4 h. MTT solution was then removed, 150 µl DMSO were added, and the absorbance of each well was measured with a microplate reader (BMG LabTech, Germany) at 570 nm. The EC_{100} values were estimated by the Graph pad Instat software as described above.

○ *In vitro* **HCV and HBV infection and treatment**

The viral host cells, human PBMCs ($1x10^6$ cells) were seeded in each well of 12-well culture plate. All wells, except negative control wells, were incubated with infected serum with either HCV (2.9×10^4 copies/ ml, genotype 4a) or HBV (1×10^4 copies/ml, genotype D) in RPMI-1640 medium for 2 h in CO_2 incubator (New Brunswick Scientific, Netherlands) at 37°C, 5% CO_2 and 95% humidity. Then the infected medium was replaced with a fresh RPMI-1640 medium containing 10% FBS for positive control wells. For treated wells, this infected medium was exchanged with RPMI-1640 medium containing 10% FBS and EC_{100} of extracts. After 96 h, untreated and treated infected cells were quantitative analyzed for the intracellular HCV and HBV. All steps were automated including the extraction of RNA or DNA on the Cobas AmpliPrep instrument and simultaneously, the PCR amplification and detection on the Cobas TaqMan analyzer (CAP-CTM). The procedure was done following the manufacturer's instructions **(Mosmann, 1983)**.

○ **Statistical analysis**

All values were expressed as mean ± standard error of the mean (SEM). Statistical significance (*p*-values< 0.01) was calculated by the multiple comparisons one-way analysis of variance (ANOVA) with post-hoc Tukey's test using SPSS16 software program.

- *In vitro* **inhibition of albumin denaturation assay**

The anti-inflammatory activity was studied using the inhibition of albumin denaturation technique **(Mizushima & Kobayashi, 1968, Sakat et al., 2010)** with minor modifications. 2.25 ml of bovine serum albumin (BSA) (1%, w/v aqueous solution) were added to 25 ml of the samples (the pigment and biosurfactant extracts solutions). Blank and standard solution (Diclofenac sodium

"Voltaren® ampoule Novartis Pharma" 1000µg/ml) were adjusted to pH 5.5 using a small amount of 0.1 N HCl, and/ or 0.1 N NaOH. The samples were incubated at 37°C for 20 min and then transferred to 60°C water bath for 5 min. Following incubation, the samples were left to cool down at room temperature then 2.5 ml of phosphate buffer was added to the above solutions. The absorbance of the above solutions was measured using UV-Visible spectrophotometer at 660 nm (UNICO-UV Visible Spectrophotometer Model UV-2000 USA). The percentage inhibition of protein denaturation was calculated using the following formula:

Inhibition of protein denaturation (%).

$$= \left(\frac{\text{absorption of control - absorption of test}}{\text{absorption of control}} \right) X\ 100$$

The control represents no inhibition of protein denaturation.

The activity of each tested supernatant was compared with the standard commercial anti-inflammatory agent "Diclofenac sodium".

- **Anti-bacterial and anti-fungal effect of the biosurfactant and pigment**

The biosurfactant and pigment of *Natrialba* sp. M6 were assessed for their ability to produce antibacterial and antifungal agents to inhibit the pathogenic bacterial and fungal indicators.

The bacterial indicators used in the current investigation were; *Bacillus subtilis* ATCC 6633, *Klebsiella pneumoniae* ATCC 13883, *Pseudomonas aeruginosa* ATCC 9027, *Escherichia coli* NCTC 10418, *Staphylococcus aureus* ATCC 6538, *Vibrio cholera* ATCC 55188. The fungal indicator strains used in the current investigation were; *Candida albicans* ATCC 10231, *Aspergillus flavus* ATCC 77708, *Rhizoctonia solani* ATCC 10532, *Fusarium solani* ATCC 11233, kindly provided by the Microbiology Lab, National Institute of Oceanography and Fisheries (NIOF), Alexandria, Egypt. All bacterial strains were maintained on nutrient agar slants, but fungal strains were maintained on PDA media incubated at 30°C. Each bacterial biomass was prepared by inoculating 100 ml of nutrient broth medium. Bacterial cultures were shaken (200 rpm) at 30°C for

24 h. Different bacterial pathogens inocula were used at the late logarithmic phase of growth ($A_{550} = 0.1$).

o **Agar well-cut diffusion assay**

Each extract was tested to contain antibacterial agents using well-cut diffusion technique, in which nutrient agar medium inoculated with indicators bacteria were poured into the plates. After solidifying, wells were punched out using 0.7 cm cork-borer and each of their bottoms was then sealed with two drops of sterile water agar. 100µl of the tested extracts was pipetted into each well. All plates were incubated at 37°C for 24 h After the incubation period, the radius of clear zones around each well (Y) and the radius of the well (X) were linearly measured in millimeters (mm). The absolute activity unit (AU) of each crude extract was calculated according to the following equation:

$$AU = Y2/ X2$$

Since, (Y) is the radius of clear zone around each well, and (X) is the radius of the well itself **(Valgas et al., 2007)**.

The fungal biomass was spot-inoculated on the fungal media containing the tested biosurfactant and pigment. Control fungal plates without biosurfactant and pigment were used to measure the decreasing in fungal growth that indicates the fungal activity of the product under investigation **(Hasan et al., 2009)**.

- **Determination of antioxidant capacity**

The pigment (1 mg) were dissolved in 1 ml dimethyl sulfoxide, and also the biosurfactant (1 mg) were dissolved in 0.1 M sodium bicarbonate (1 ml) to reach a final concentration of 1000 µg/ml from each tested compound. The antioxidant activity of the pigment and the biosurfactant solutions (200 µl) were investigated using phosphomolybdenum reagent (32 mM sodium phosphate, 4 mM ammonium molybdate, and 0.6 M sulfuric acid) according to **(Sikkandar et al., 2013).** Finally, the absorbance was measured at 695 nm, and the antioxidant capacity was expressed as standard butylhydroxytoluene equivalents (BHT).

- **Anti-biofouling activity**

To study the biofouling inhibition, in 250 ml conical flasks, 100 ml of nutrient broth containing glass slide were inoculated with

1 ml of sea water was incubated at 28°C for 24 h. After that, 1000 µg/ml of the tested pigment or biosurfactant were added. A control flask was prepared using the same conditions without adding pigment or biosurfactant. After incubation period the formed biofilms adhered to the surface of the glass slides were stained with 0.4 % crystal violet solution for 10 min, then washed with water, air-dried and were finally observed under the microscope **(Bavya et al., 2011, Kumar & Aparna, 2014).**

Results

Relation between the growth and the biosurfactant or pigment production

Based on previous data, it was planned in this experiment to compare growth, biosurfactant activity and pigment production of *Natrialba* sp. M6 on basal and pre-optimized media. In the basal condition, M5 contained 3.4 M NaCl, pH 11 and cultures were incubated at 37°C, whereas in the pre-optimized, the bacteria were grown in M5 amended with 2.57 M NaCl, adjusted to pH 10 and incubated at 45°C.

The results in **Fig. (2)** showed that, with respect to growth expressed as dry weight, the basal medium supported better growth **(Fig. 2a).** Biosurfactant activity showed higher values in basal medium compared to pre-optimized medium **(Fig. 2b).** The highest value 26.31(1/ST*1000) was attained in 12 days old cultures. On the contrary, the pigment production was better in pre-optimized medium compared to basal medium **(Fig. 2c).** The maximum production of the pigment (2.5 mg) was recorded by 11 days old cultures.

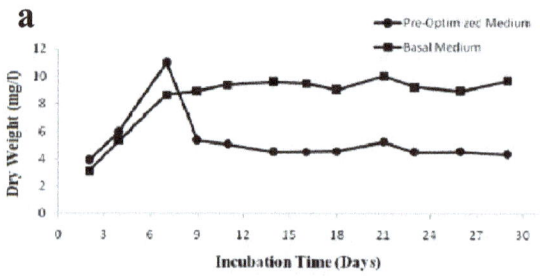

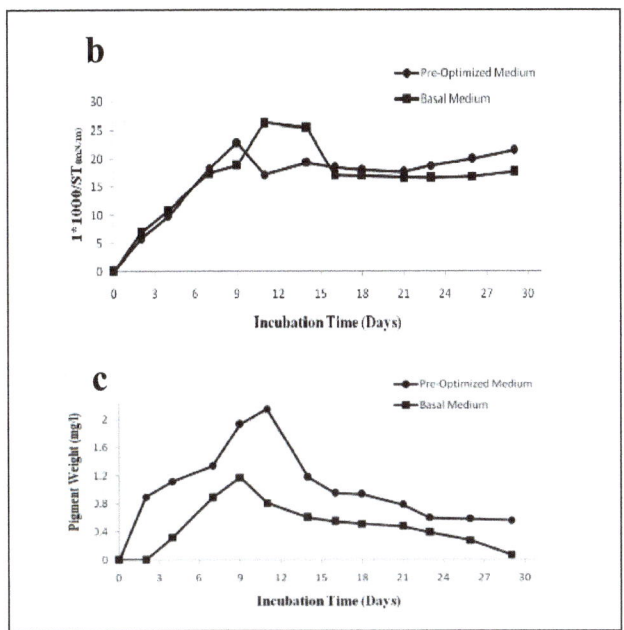

Fig. 2: Growth as dry weight (a), biosurfactant production (b), and pigment formation (c) at both basal and pre-optimized media for *Natrilba* sp. M6.

Optimization of *Natrialba* sp. M6 biosurfactant using experimental designs

In order to study the significance of various nutritional and culture conditions on biosurfactant production by the tested bacterium, Plackett-Burman Design was applied. For this screening step, 14 variables at two different levels, high and low were selected. **Tables (1&2)** present the design matrix and the response of the different trials (the reciprocal of the corrected surface tension * 1000). The main effect of each variable on the surface tension was estimated as the difference between the average of the measurements at the high (1) and low (-1) levels of that factor. The data shown in **Table (2)** illustrate a variation in the response from 14 to 22.

53

Table 1: Variables and their levels employed in Plackett-Burman Design for screening of culture conditions affecting the biosurfactant production by *Natrialba* sp. M6.

Variable symbol	Variable name	Low level (-1)	High level (+1)
X1	Temperature	35°C	45°C
X2	pH	8	10
X3	Casaminoacids	0.50 (g%)	1.00 (g%)
X4	NaCl	10.00 (g%)	20.00 (g%)
X5	Agitation	Static	200.00 rpm
X6	Glucose	0.10 (g%)	1.00 (g%)
X7	Glycerol	0.10 (g%)	1.00 (g%)
X8	NH_4Cl	0.01 (g%)	0.10 (g%)
X9	NH_4NO_3	0.01 (g%)	0.10 (g%)
X10	Yeast Extract	0.10 (g%)	1.00 (g%)
X11	$(NH_4)_2SO_4$	0.01 (g%)	0.10 (g%)
X12	$MgSO_4.7H_2O$	0.01 (g%)	0.10 (g%)
X13	$CaCl_2$	0	0.10 (g%)
X14	$FeSO_4$	0	0.01 (g%)

Table 2: Plackett-Burman experimental design for the evaluation of the factors influencing biosurfactant activity of *Natrialba* sp. M6

Trials	X_1	X_2	X_3	X_4	X_5	X_6	X_7	X_8	X_9	X_{10}	X_{11}	X_{12}	X_{13}	X_{14}	Surface Tension mN/m	Response* 1/ST*1000
1	-1	1	1	-1	1	-1	1	-1	1	1	1	-1	1	-1	56.5	17.699
2	1	-1	-1	-1	1	-1	-1	-1	-1	1	1	-1	1	1	55.9	17.889
3	-1	-1	-1	-1	1	1	1	1	1	-1	-1	-1	1	1	67.5	14.815
4	1	1	1	1	1	1	1	-1	1	-1	1	1	1	-1	57.9	17.271
5	1	-1	1	1	1	1	-1	1	1	1	-1	1	1	-1	60.5	16.529
6	-1	-1	1	1	1	-1	-1	-1	-1	-1	-1	1	-1	1	64.3	15.552
7	1	1	-1	1	-1	-1	1	-1	1	1	-1	1	-1	1	65.9	15.175
8	1	1	1	1	1	-1	-1	1	1	-1	-1	1	1	1	62.2	16.077
9	1	1	1	1	1	-1	1	1	1	-1	-1	1	-1	1	43.6	22.936
10	1	-1	1	-1	-1	1	-1	-1	-1	-1	1	-1	-1	-1	64.6	15.48
11	1	1	1	-1	-1	1	1	1	-1	-1	1	1	1	1	65.2	15.337
12	-1	-1	1	1	-1	1	-1	1	-1	-1	-1	1	1	-1	62.3	16.051
13	-1	1	-1	-1	1	1	-1	-1	-1	1	1	-1	1	-1	66.2	15.106
14	-1	1	1	-1	-1	1	-1	1	1	1	-1	-1	-1	-1	68.5	14.599
15	-1	1	1	1	-1	-1	-1	-1	1	1	-1	-1	-1	1	65.1	15.361
16	-1	-1	-1	-1	-1	-1	1	1	1	1	1	1	1	1	58.3	17.153

Response is the reciprocal of the measured surface tension * 1000

Statistical analysis of the Plackett-Burman Design

Since Plackett-Burman design is two levels experimental design, it involves a linear polynomial correlation model that describes the correlation between the 14 factors and the response as follow:

Y= 16.907 + 0.35 X1+ 0.371 X2 + 0.292 X3 + 0.502 X4 + 0.956 X5 - 0.46 X6 + 0.303 X7 + 0.062 X8 - 0.74 X9 - 0.42 X10 - 0.38 X11- 0.4 X12 - 0.19 X13 -0.00515 X14

As shown from the data of the *p*-value in (**Table 3**), the most significant factors affecting biosurfactant production are agitation, ammonium nitrate, sodium chloride, glucose and yeast extract in descending order. Although, ammonium nitrate, glucose, yeast extract, ammonium sulfate and magnesium sulfate more significantly affect the biosurfactant production than pH and glycerol, they were omitted and glycerol was selected for subsequent optimization steps. The reason was that all ammonium nitrate, glucose, yeast extract, ammonium sulfate and magnesium sulfate showed negative main effect and the used low value (-1) for all in plackett-Burman design was zero, that could not be further decreased in the next optimization step. Therefore, pH, glycerol, agitation and sodium chloride were selected as significant factors for further optimization and these results were checked in the pre-optimization step.

When analyzing the regression coefficient for the 14 variables, it is concluded that pH, agitation, glycerol, temperature, casamine acids, NH_4Cl and NaCl showed positive effect on the biosurfactant activity. On the other hand, all other variables showed negative effect on the biosurfactant production (**Fig. 3**).

The main effects of the examined factors on biosurfactant production were calculated and presented in **Table (3)** and **Fig. (3)**.

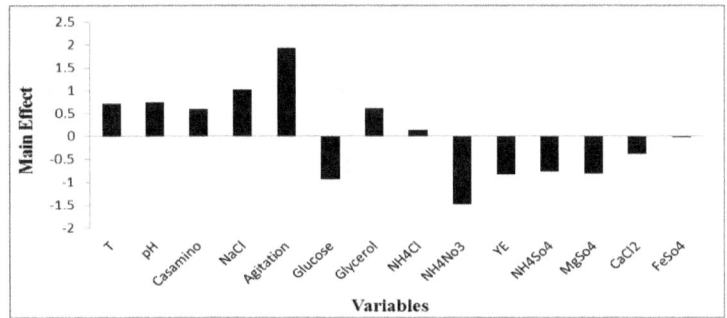

Fig. 3: Main effect of the different factors influencing the biosurfactant production by *Natrialba* sp. M6 based on Plackett-Burman Design.

Pareto chart presented in **Fig (4)** is another convenient way of presenting the results, which display the magnitude of each estimate, and thereby showing the ranking of the factor estimates. As shown in the figure, the most significant factors affecting biosurfactant activity were agitation, ammonium nitrate, sodium chloride, glucose and yeast extract in descending order. Agitation and sodium chloride were selected as significant factors for further optimization since ammonium nitrate, glucose and yeast extract had negative effect and their low value equal to zero.

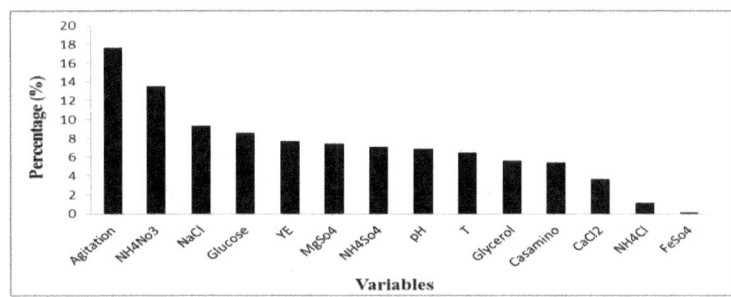

Fig. 4: Pareto chart rationalizing the effect of each factor on the biosurfactant production by *Natrialba* sp. M6.

Table 3: Statistical analysis of Plackett-Burman Design experiment for biosurfactant production by *Natrialba* sp. M6.

Variables	Coefficients	Main Effect	t Stat	*P*-value
Intercept	16.90686	0.7004	35.25068	0.000804
T	0.3502	0.741014	0.708364	0.552151
pH	0.370507	0.583268	0.749438	0.531753
Casaminoacids	0.291634	1.004412	0.5899	0.615026
NaCl	0.502206	1.912744	1.015833	0.416604
Agitation	0.956372	-0.92458	1.934492	0.192718
Glucose	-0.46229	0.606822	-0.9351	0.448451
Glycerol	0.303411	0.125328	0.613722	0.601903
NH_4Cl	0.062664	-1.47008	0.126754	0.910729
`NH_4NO_3	-0.73504	-0.83042	-1.48679	0.275429
Yeast extract	-0.41521	-0.76582	-0.83986	0.489383
$(NH_4)_2SO_4$	-0.38291	-0.80344	-0.77453	0.519647
$MgSO_4.7H_2O$	-0.40172	-0.3896	-0.81257	0.501805
$CaCl_2$	-0.1948	-0.01232	-0.39403	0.7316
$FeSO_4$	-0.00616	0.7004	-0.01246	0.991192

Optimization of the culture conditions using Box-Behnken Design

To identify the optimum response region for biosurfactant production, the significant independent variables (pH, X_1; glycerol, X_2; agitation, X_3; sodium chloride, X_4) were further explored at three levels based on the results and the two levels used in Plackett-Burman Design illustrated in **(Table 4)**. For example, the results of Plackett-Burman showed that pH has positively affect the biosurfactant production, therefore, it was explored in three levels in Box-Behnken design considering the high level (1) in Plackett-Burman to be the low value (-1) in Box-Behnken, then two higher values were selected encoding zero as a medium level and the high level values (1).

Table (5) presents the design matrix for the variables and the response of each trial, where the response here was the reciprocal of the corrected surface tension * 1000 (1/ST * 1000).

To predict the optimal point, a second order polynomial function was fitted to the experimental response results (non-linear optimization algorithm).

$Y= 22.69136 + 2.311071X_1 + 1.297846X_2 - 3.40958X_3 + 0.191705X_4 + 0.661656X_1X_2 - 6.70802X_1X_3 + 0.324812X_1X_4 - 8.66472X_2X_3 + 0.237175X_2X_4 - 0.22231X_3X_4 + 0.490423X_1^2 - 0.84625X_2^2 + 4.732054X_3^2 - 3.00752X_4^2$

On the model level, the correlation measures for estimating the regression equation are the multiple correlation coefficients R and the determination coefficient R^2. In this experiment, the value of R^2 was 0.738 for the biosurfactant production, indicating a high degree of correlation between the experimental and the predicted values.

The optimal levels of the three factors, as obtained from the maximum point of the polynomial model, were estimated using the *solver* function of the Microsoft Excel tools and found to be (1) pH equal to 12, (1) 3% glycerol, (-1) agitation equal to 150 rpm and (0.165583) 20.8% NaCl with the predicted response (the reciprocal of surface tension * 100) equal to 50.26165.

In addition, **Fig. (5)** shows the simultaneous effects of the three most significant independent factors on each response using three-dimensional graphs generated by Statistica 5.0 software

For graph a, the graph showed foci for maximum level of biosurfactant production at high values of both glycerol and pH. However, graph b showed the maximum response at high pH and low agitation. Meanwhile, in graph c, it was indicated that at middle NaCl and high pH values the maximum production was achieved, but at high glycerol value and low agitation the maximum biosurfactant production was obtained in graph d, graph e showed that at low agitation value and middle value of NaCl gave the highest biosurfactant production level.

From the above presented data, it could be concluded that the amount of biosurfactant has increased from 22.20 to 49.02 indicating that the amount of biosurfactant has been doubled (220%) due to the optimization process.

Verification of the model

The optimal conditions realized from the optimization experiment were verified experimentally and compared with the data calculated from the model. The estimated response was 48.8, whereas the predicted value from the polynomial model was 50.2, thereby confirming the high accuracy (97.2%) of the model under the investigated conditions.

The optimum formula for the maximum biosurfactant production by *Natrialba* sp. M6 each separately, finally was as follows:

Components	Basal Medium	Optimized medium
Temperature	37°C	45°C
pH	11	12
Casaminoacids	0.5	1
NaCl	20	20.8
Agitation	200	150rpm
Glucose	0	0.1
Glycerol	0	3
NH_4Cl	0	0.1
NH_4NO_3	0	0.01
Yeast Extract	0	0.01
$(NH_4)_2SO_4$	0	0.01
$MgSO_4.7H_2O$	0.02	0.01
$CaCl_2$	0	0
$FeSO_4$	Traces	0.01
Volume in ml	100	100
Biosurfactant $(mN/m)^{-1}$	22.2	49.02

*Glucose was added after autoclaving as sterile solution in a known calculated volume.

Table 4: The three levels of significant independent variables used in Box-Behnken factorial experimental design for biosurfactant production by *Natrialba* sp. M6

Level	X1: pH	X2: Glycerol (g %)	X3: Agitation (rpm)	X4: NaCl (g %)
1	12	3	250	25
0	11	2	200	20
-1	10	1	150	15

Table 5: Box-Behnken factorial experimental design for biosurfactant production by *Natrialba* sp. M6.

Trial	pH	Glycerol	Agitation	NaCl	Response* $(1/ST * 1000)$
1	0	0	0	0	24.69
2	0	0	1	-1	20.37
3	-1	-1	0	0	22.52
4	0	0	1	1	19.92
5	0	0	-1	1	23.75
6	1	-1	0	0	17.70
7	0	0	-1	-1	23.31
8	-1	1	0	0	20.49
9	1	1	0	0	18.32
10	0	0	0	0	18.12
11	0	1	1	0	20.70
12	0	-1	1	0	31.35
13	0	1	-1	0	42.92
14	1	0	0	-1	21.93
15	-1	0	0	1	21.55
16	0	-1	-1	0	18.90
17	-1	0	0	-1	20.45
18	1	0	0	1	24.33
19	0	-1	0	-1	19.16
20	0	1	0	1	20.37
21	0	1	0	-1	20.50

Trial	pH	Glycerol	Agitation	NaCl	Response* $(1/ST * 1000)$
22	-1	0	-1	0	20.37
23	0	0	0	0	20.45
24	1	0	1	0	23.42
25	1	0	-1	0	49.02
26	0	-1	0	1	18.08
27	-1	0	1	0	21.60

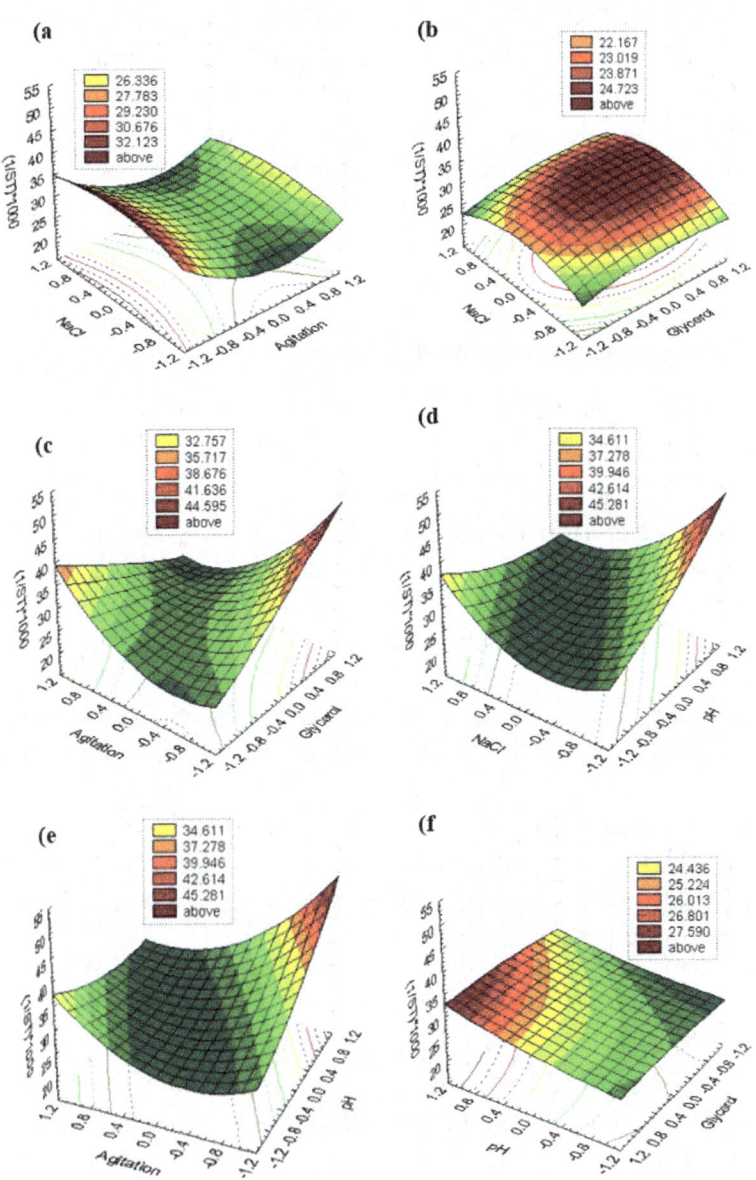

Fig. 5: Three-dimensional surface and contour plots showing the relationships between the tested variables and the biosurfactant as a response in a form of [(1/ST)*1000] produced by *Natrialba* sp. M6. (a) Showed that low agitation

value and middle value of NaCl gave the highest biosurfactant production level. (**b**) Showed that high glycerol value and middle value of NaCl gave the highest biosurfactant production level. (**c**) Showed foci for maximum level of biosurfactant production at high glycerol value and low agitation. (**d**) Indicated that at middle NaCl and high pH values the maximum production was achieved. (**e**) Showed the maximum response at high pH and low agitation and (**f**) showed foci for maximum level of biosurfactant production at high values of both glycerol and pH.

Optimization of pigment produced by *Natrialba* sp. M6 using Box-Behnken design

The most significant independent variables from the previous experiment (Sodium chloride, X_1; pH, X_2; and Time (culture age), X_3) were further explored at three levels illustrated in **Table (6). Table (7)** presents the design matrix for the variables and the response of each trial, where the response here was the concentration of pigment µg/ml after measuring the absorbance of the pigment at wavelength 350 nm. In this experiment, the value of R^2 was 0.813 for the pigment formation, indicating a high degree of correlation between the experimental and the predicted values.

To predict the optimal point, a second order polynomial function was fitted to the experimental response results (non-linear optimization algorithm).
$Y = 7912.805 + 1597.5336X_1 - 371.9934X_2 - 17.835X_3 + 783.0545X_1X_2 + 1491.3712X_1X3 + 1262.0602X_2X_3 - 1216.764X_1^2 - 1452.87X_2^2 - 496.5573X_3^2$.

The optimal levels of three factors, as obtained from the maximum point of the polynomial model, were found to be NaCl 25%, pH equal to 10.07, the culture age equal to 11 days with the predicted concentration equal to 9.8227 mg/ml equivalent to λ_{350} absorbance 1.453506. **Fig. (6)** shows the simultaneous effects of three most significant independent factors on each response using three-dimensional graphs generated by Statistica 5.0 software.

The data presented in graph (a) showed foci for maximum level of carotenoid synthesis, indicating a great interaction between pH and NaCl at high values. However, the data illustrated in graph (b) indicate that the graph was high facing the culture age axis with maximum response around its high values. This indicating that high NaCl and culture age were more preferable for higher carotenoid synthesis for *Natrialba* sp. M6 in graph (c), it was completely directed against pH axis, indicating that culture age with the same middle values of pH gave the maximum carotenoid yield.

Verification of the model

The estimated pigment was 9.992 mg/L (λ_{350} absorbance 1.47963), whereas the predicted value from the polynomial model was 9.8226 mg/L (absorbance λ_{350} 1.453506). This means the high accuracy (~92.87%) of the model under the predicted conditions.

The optimum formula for the maximum pigment production by *Natrialba* sp. M6 each separately, finally was as follows:

Components	Basal Medium	Optimized medium
Temperature	45°C	45°C
pH	11	10
Casaminoacids (g)	0.5	0.5
NaCl (g)	15	25
Na_2CO_3 (g)	0.9	0.9
Agitation	200	200rpm
$MgSO_4.7H_2O$ (g)	0.02	0.02
KH_2PO_4 (g)	0.1	0.1
Trace element solution (ml)	0.1	0.1
Culture age (Days)	7	9
Volume in ml	100	100
Pigment (mg/L)	2.3	9.8226

Table 6: The three levels of significant independent variables used in Box-Behnken factorial experimental design for pigment formation by *Natrialba* sp. M6.

Level	X1: NaCl (%)	X2: pH	X3: Inoculum age (days)
1	25	11	11
0	20	10	9
-1	15	9	7

Table 7: Box-Behnken factorial experimental design for pigment formation by *Natrialba* sp. M6.

Trial	NaCl	pH	Inoculum age	Response* Abs. λ_{350}nm	Pigment conc. mg/ml
1	0	0	0	1.176	7.87
2	0	0	0	1.183	7.92
3	0	-1	1	0.59	3.89
4	0	1	-1	0.83	5.52
5	-1	-1	0	0.98	6.53
6	0	0	0	1.189	7.95
7	1	0	-1	1.032	6.89
8	0	1	1	1.152	7.70
9	0	-1	-1	1.011	6.77
10	-1	0	1	0.39	2.53
11	-1	0	-1	0.79	5.24
12	1	0	1	1.51	10.14
13	1	1	0	0.83	5.52
14	1	-1	0	1.009	6.73
15	-1	1	0	0.34	2.19

*Response is the absorbance of the pigment after extraction measured at wavelength 350 nm, then the related concentration was measured as µg/ml in the second column based on correlation equation included in the material section.

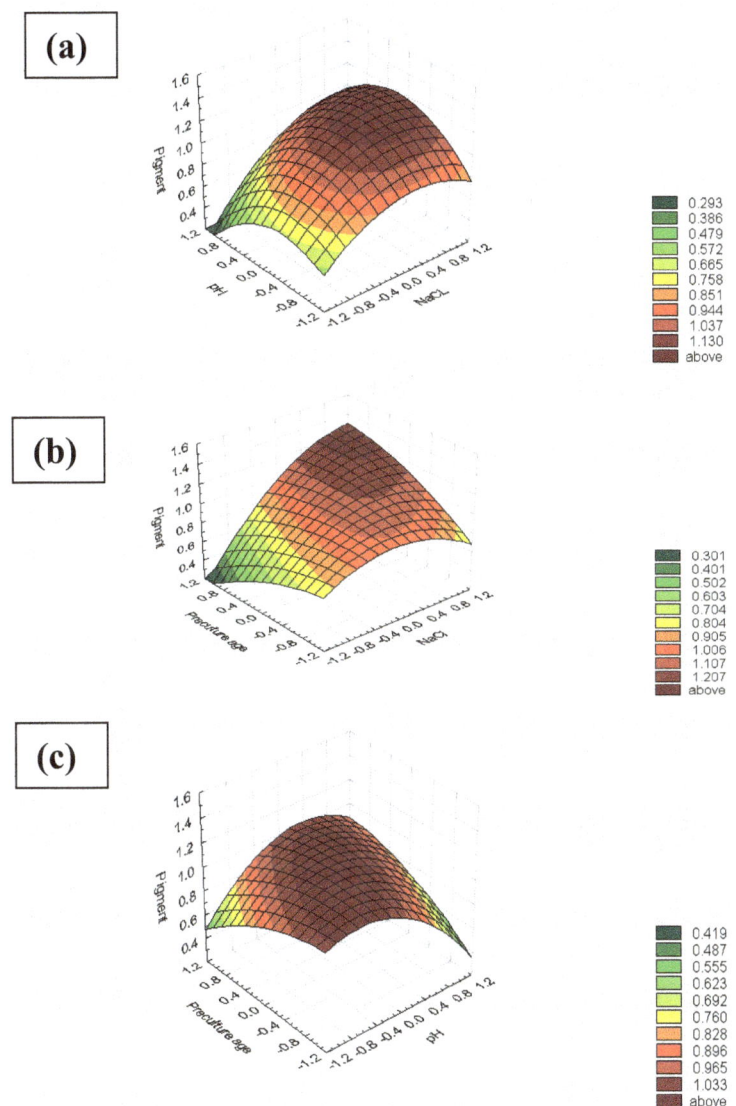

Fig. 6: Three-dimensional surface plots showing the relationships between tested significant variables and the pigment formation by *Natrialba* sp. M6, the optimal levels of the three factors, as obtained from the maximum point of the polynomial model.

Identification of partially purified *Natrialba* sp. M6 biosurfactant

- **Gas chromatography-mass spectrophotometer (GC-MS) analysis**

The GC-MS analysis of the partially extracted biosurfactant, as shown in **Fig. (7)**, indicates the presence of different biosurfactant components. Detailed description for the existed major peaks of the active compounds was shown in **Table (8)**. The structure of the major surfactants components present in the biosurfactant extract is summarized in **Table (9)**.

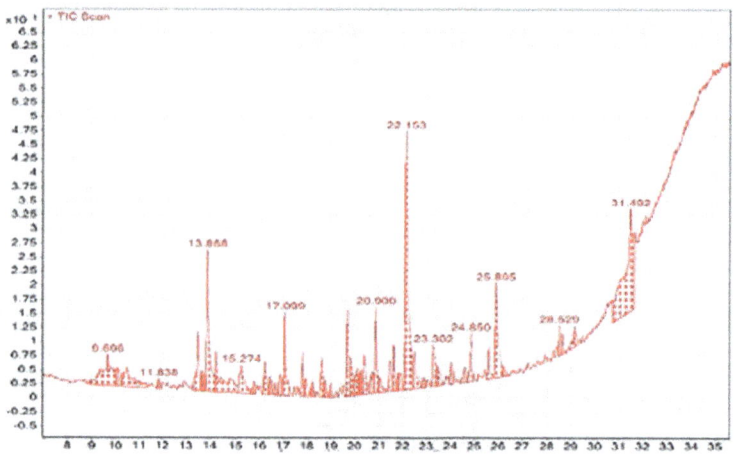

Fig. 7: GC-MS analysis of partially extracted biosurfactant.

Table 8: Peak integration list

Peak	Start	RT	End	Height	Area	Area%
1	13.288	13.513	13.611	483266.59	3204323.92	5.79
2	13.825	13.888	14.212	1157543.27	8416056.82	15.2
3	17.036	17.099	17.284	684887.82	3152193.26	5.69

4	19.585	19.693	19.797	708554.39	4637574.07	8.38
5	20.819	20.9	21.009	704675.61	3307443.82	5.97
6	21.604	21.656	21.743	371353.71	1526066.31	2.76
7	22.06	22.153	22.465	2111472.47	19200516.41	34.68
8	24.694	24.85	24.942	396021.36	2006770.35	3.62
9	25.803	25.895	26.167	774621.6	6773550.25	12.23
10	30.741	31.492	31.63	835057.56	19453966.55	35.13

Table 9: The major components present in the biosurfactant extract of *Natrialba* sp. M6 identified by gas chromatography mass spectrophotometer (GC-MS) analysis.

No	Compounds	Chemical structure	Molecular formula (M.wt)	RT (min)	Prob. %
1	Octadecane,1-[2-(hexadecyloxy)ethoxy]		$C_{36}H_{74}O_2$ (538.97)	9.696	5.30
2	Ethanol, 2-(octadecyloxy)		$C_{20}H_{42}O_2$ (314.55)	13.525	6.64
3	Phenol, 2,4-bis(1,1-dimethylethyl)		$C_{14}H_{22}O$ (206.32)	13.894	51.1
4	Tetrapentacontane, 1,54-dibromo		$C_{54}H_{108}Br_2$ (917.24)	14.264	2.45
5	1,3-Dioxane,5-(hexadecyloxy)-2-pentadecyl-, trans		$C_{35}H_{70}O_3$ 538.93	15.292	7.45

No	Compounds	Chemical structure	Molecular formula (M.wt)	RT (min)	Prob. %
6	Ethyl iso-allocholate		$C_{26}H_{44}O_5$ (436)	16.245	50.8
7	Tert-Hexadecanethiol		$C_{16}H_{34}S$ 258.51	18.641	15
8	Pentadecanoic acid, ethyl ester		$C_{17}H_{34}O_2$ 270.45	19.716	44.1
9	Octadecanoic acid, 4-hydroxy-, methyl ester		$C_{19}H_{38}O_3$ (314.5)	20.414	4.10

No	Compounds	Chemical structure	Molecular formula (M.wt)	RT (min)	Prob. %
10	Ethyl 14-methyl-hexadecanoate		$C_{19}H_{38}O_2$ (298.5)	21.500	3.92
11	14-Hydroxy-15-methylhexadec-15-enoic acid, ethyl ester		$C_{19}H_{36}O_3$ (312.49)	21.500	3.08
12	Eicosanoic acid, ethyl ester		$C_{22}H_{44}O_2$ (340.58)	22.188	7.51
13	Ethyl tridecanoate		$C_{15}H_{30}O_2$ (242.4)	22.188	6.06

No	Compounds	Chemical structure	Molecular formula (M.wt)	RT (min)	Prob. %
14	9,12,15-Octadecatrienoic acid, 2-[(trimethylsilyl)oxy]-1-[[(trimethylsilyl)oxy]methyl]ethyl ester, (Z,Z,Z)		$C_{27}H_{52}O_4S_{i_2}$ (496.87)	22.517	10.1
15	Ethyl stearate, mono 9-epoxy		$C_{20}H_{37}O_3^-$ (325.51)	23.302	1.13

- ## Percentage composition (dry weight)

The study presented the measurement of the percentage composition of the biosurfactant sample, which included protein, carbohydrate and lipid contents of *Natrialba* sp. M6 biosurfactant was shown in **Table (10).**

Table 10: Percentage composition of protein, carbohydrates and lipid of *Natrialba* sp. M6 biosurfactant.

Sample	Protein	Carbohydrate	Lipid
Biosurfactant	31.599±0.045	11.71875±0.022	41.53±0.021

Partial characterization of the pigment from *Natrialba* sp. M6

In this part of the work, it was planned to characterize the pigment produced using two techniques: FT-IR and Raman spectroscopy.

- ## Fourier-Transform Infrared Spectroscopy (FT-IR)

The results of FT-IR spectroscopic studies revealed the presence of various chemical groups in the pigment extract produced by *Natrialba* sp. M6 **(Fig. 8)**. The peaks at 543.94 and 1020.38 cm^{-1} correspond to C-Br and C-F stretching frequency, respectively. A band at 1134.18 cm^{-1} corresponds to C-OH. The peak at 1263.42 cm^{-1} is assigned to C-O-C. The peaks at 1408.08 and 1454.38 cm^{-1} correspond to CH$_3$ and CH$_2$ stretching frequency, respectively. The strong peak at 1637.62 cm^{-1} is assigned to the C=C alkene stretching which means that some aliphatic compounds existed in this pigment extract.

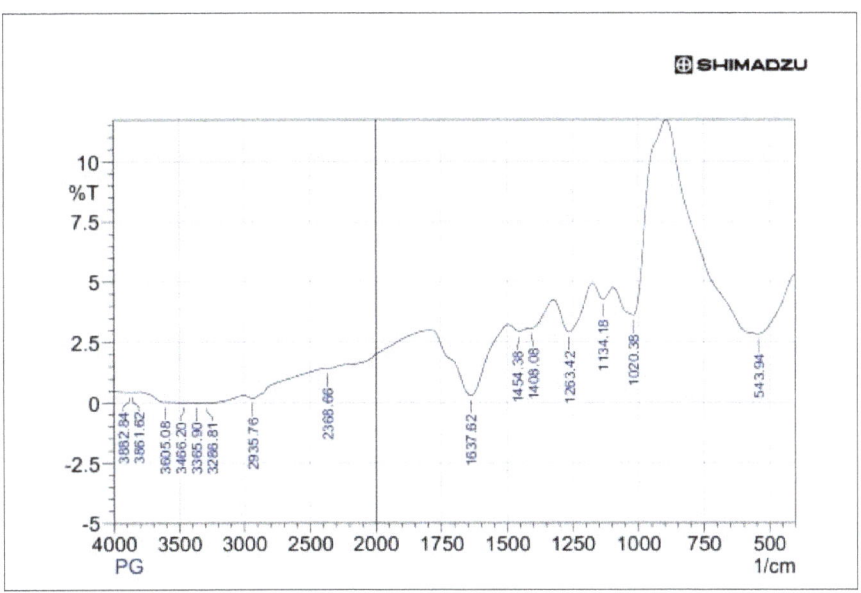

Fig. 8: Fourier-Transform Infrared spectrum of the pigment.

- **Raman spectroscopy**

Results of Raman spectroscopy analysis revealed the presence of three regions with two strong and weak signal intensities **(Fig. 9)** of the orange carotenoids showing strong band at 1380 cm^{-1}, which corresponds to CH$_3$ umbrella mode (weak band at 1883 cm^{-1}) corresponds to C=C bond.

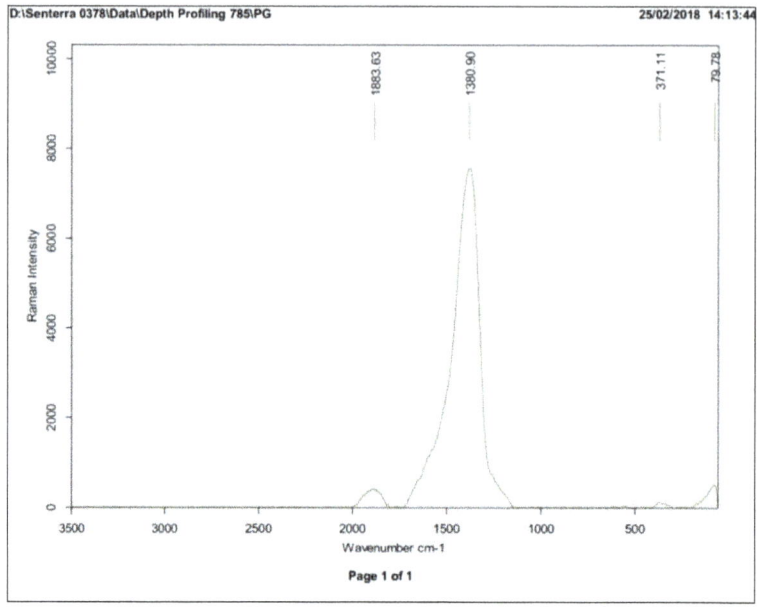

Fig. 9: Raman spectra of the pigment signal.

Biological activity of *Natrialba* sp. M6 biosurfactant and pigment

- Cytotoxicity on normal human lung fibroblast cells and human cancer cell lines

The cytotoxic effects of different concentrations of biosurfactant and pigment on normal human lung fibroblast cells (Wi-38), Caco-2 (colon cancer line), HepG-2 (liver cancer line), MCF-7 (breast cancer cell line), and Hela (cervical cancer cell line) was achieved by trypan blue dye exclusion assay. As shown in **Tables (11)** and **(12)**, treatment of normal cells and cancer cell lines with different concentrations of either biosurfactant or pigment extraction showed cytotoxic effects on the tested cell lines. The high IC_{50} and EC_{100} values are indicators for safety of the extracts toward normal cells. This means that the highest IC_{50} and EC_{100} values of pigment proved its highest safety compared to biosurfactant and the currently used chemotherapy (FU) for the cancer cell lines, while, the low IC_{50} value is indicator for high anticancer activity. These results indicate that the low IC_{50} of FU and pigment proved its high anticancer potential against different cancer cells compared to biosurfactant. Morphological changes of human cancer cell lines; Caco-2, MCF-7, HepG-2 and Hela as well as human normal cell line (Wi-38) before and after 72 h treatment with either pigment or biosurfactant and currently used chemotherapy (FU) are shown in **Fig. (10)**.

Table 11: IC_{50} ($\mu g/ml$) and EC_{100} ($\mu g/ml$) of biosurfactant and pigment toward human normal cell line (Wi-38).

Samples	Wi-38	
	IC_{50}	EC_{100}
Pigment	85.14 ± 2.31^{a}	31.6 ± 0.96^{a}
Biosurfactant	14.75 ± 1.09^{b}	6.85 ± 0.52^{b}
FU	10.28 ± 0.67^{b}	3.3 ± 0.48^{b}

All values are expressed as mean±SE. Different letters are significantly different within the same column at p<0.05.

Table 12: IC_{50} ($\mu g/ml$) of biosurfactant and pigment against human cancer cell lines

IC_{50} ($\mu g/ml$)	Caco-2	MCF-7	HepG-2	Hela
Pigment	34.62 $\pm1.07^{b}$	21.18 $\pm0.77^{b}$	38.24 $\pm0.379^{b}$	28.91 $\pm0.05^{b}$
Biosurfactant	233.57 $\pm1.5^{c}$	134.35 $\pm6.29^{c}$	239.32 $\pm0.823^{c}$	222.69 $\pm1.35^{c}$
FU	7.17 $\pm0.0001^{a}$	7.98 $\pm0.013^{a}$	10.36 $\pm0.46^{a}$	6.39 $\pm0.47^{a}$

All values are expressed as mean±SE. Different letters are significantly different within the same column at p<0.05.

- ## Cancer selectivity index (SI)

The ratio of the IC_{50} for the normal human cells and the IC_{50} for the cancer cell lines can be measured, the high SI value is indicator for high selectivity toward cancer cells **(Table 13)**. This means that the highest SI value of pigment, proved its highest selectivity between normal and cancer cells compared to currently used chemotherapy (FU) and biosurfactant.

Table 13: Cancer selectivity index (SI) of *Natrialba* sp. M6 biosurfactant and pigment.

	SI			
Samples	Caco-2	MCF-7	HepG-2	Hela
Pigment	2.46 ± 0.14^a	4.03 ± 0.25^a	2.23 ± 0.08^a	2.94 ± 0.08^a
Biosurfactant	0.063 ± 0.004^c	0.11 ± 0.003^c	0.061 ± 0.004^c	0.07 ± 0.004^c
FU	1.43 ± 0.09^b	1.29 ± 0.082^b	0.99 ± 0.02^b	1.62 ± 0.22^b

All values are expressed as mean±SE. Different letters are significantly different within the same column at $p<0.05$.

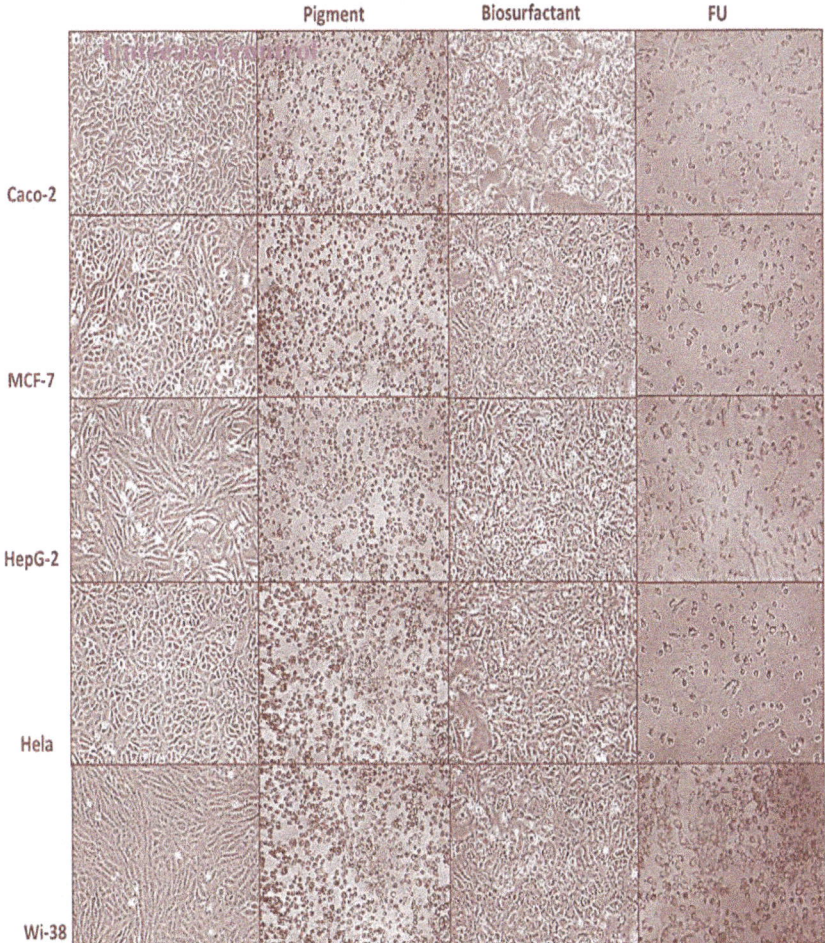

Fig. 10: Morphological changes of human cancer cell lines; Caco-2, MCF-7, HepG-2 and Hela as well as human normal cell line (Wi-38) before and after 72 h treatment with pigment, biosurfactant and currently used chemotherapy (FU). The figure depict that the effect of pigments treatment is close to the chemotherapy treatment, meanwhile, the biosurfactant treatment is not different from the untreated control.

- Flow cytometry analysis of cell death

Cancer cells were treated with the IC_{50} doses of more effective extract and FU for 72h in 5% CO_2 incubator and then stained with two different stains (Annexin V- biotin and streptavidine-fluorescein). The cell death rates were measured using flow cytometry estimation of the percentage annexin-stained apoptotic cells and propidium iodide (PI)-stained necrotic cells. The high percentage of annexin-stained cancer cells is an indicator for the mechanism of cell death through apoptosis. This means that pigment has mediated apoptosis-dependent cell death in different treated cancer cells. Moreover, **Table (14)** and **Fig. (11)** demonstrate that pigment caused apoptosis by higher percentage ($\geq$ 49%) than the currently used chemotherapy (<39%).

Table 14: The percentage of apoptotic cells treatment cancer cells with pigment and FU.

Samples	Total Apoptosis			
	Caco-2	MCF-7	HepG-2	Hela
Control	0±0	0.1±0.005	1.75±0.03	1.87±0.005
Pigment	51.48±1.61	53.55±0.47	48.87±0.84	52.38±0.05
- **FU**	38.01±0.67	32.88±0.43	33.12±0.19	38.05±0.63

All values are expressed as mean±SE. Different letters are significantly different within the same column at p<0.05.

- Detection of cell death by using fluorescence phase contrast microscope

Cancer cells were treated with IC_{50} doses of more effective extract and FU for 72 h in 5% CO_2 incubator and then were stained by two fluorescents nuclear staining using ethidium bromide and acridine orange dyes, then the cells were examined using the fluorescence phase contrast microscope. Green, orange, and red fluorescence refer to normal cells, early apoptotic cells, and late apoptotic cells, respectively **(Fig. 12)**.

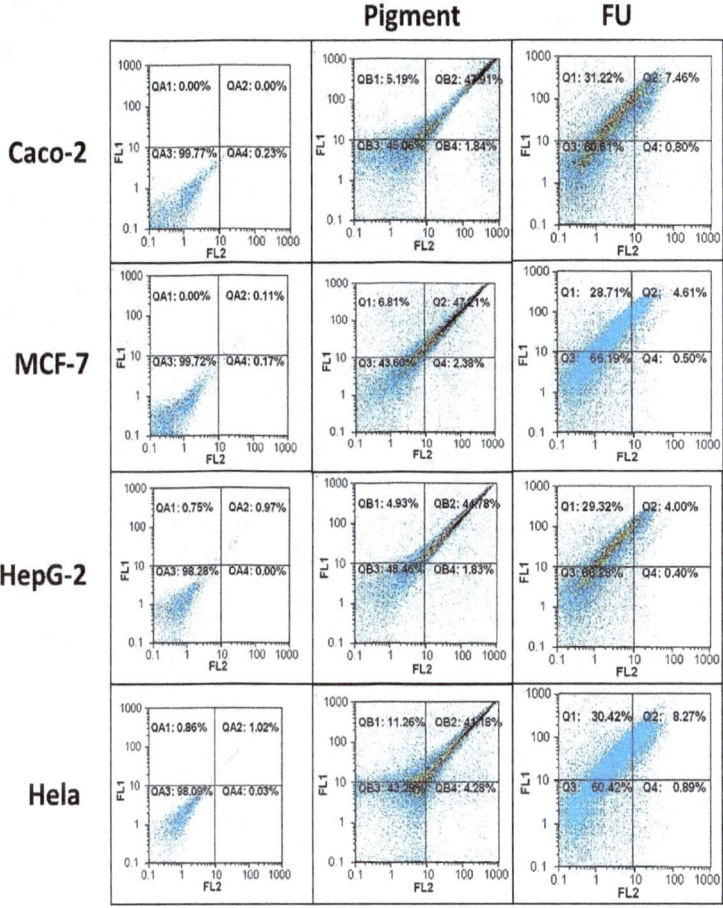

Fig. 11: Annexin-PI analysis of Caco-2, MCF-7, HepG-2, and Hela before and after 72 h treatment with pigment and currently used chemotherapy (FU). The color gradient indicates different stages of apoptotic cell death.

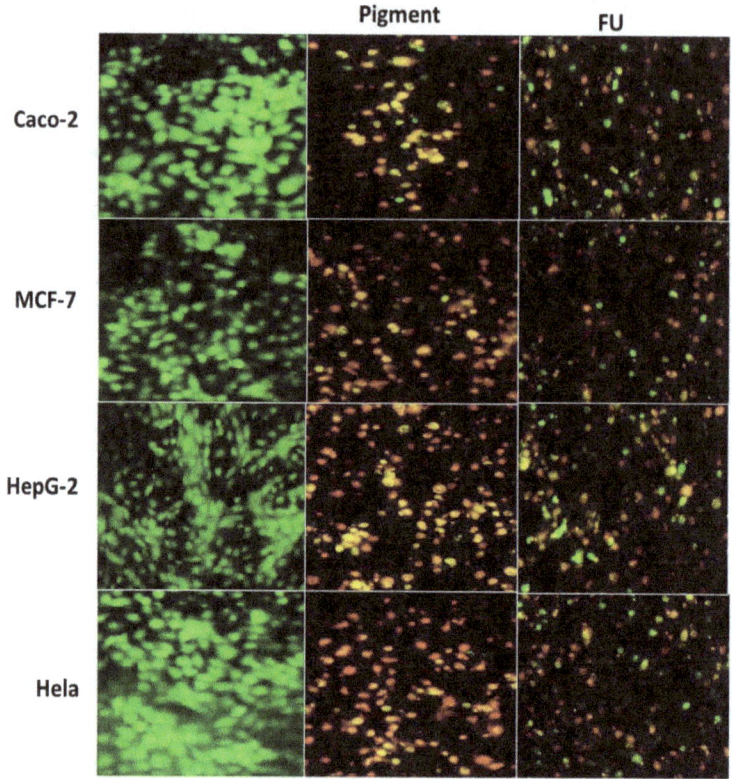

Fig. 12: Morphological observation of cancer cell lines by AO-EB double staining technique using fluorescence microscope of Caco-2, MCF-7, HepG-2, and Hela before and after 72 h treatment with pigment and currently used chemotherapy (FU). Green cells are viable cells, orange cells are the early apoptotic cells, and red cells are the late apoptotic cells.

- **Cytotoxicity on PBMCs**

The cytotoxic effects of different concentrations of biosurfactant and pigment on viral host cells, human PBMCs were studied, the high EC_{100} value are indicator for safety pigment than biosurfactant samples toward normal PBMCs cells but not more than Sovaldi (the drug used for the treatment of hepatitis C) as illustrated in **Table (15)**.

Table 15: EC$_{100}$ (µg/ml) of pigment, biosurfactant and sovaldi toward human PBMCs.

Samples	EC$_{100}$
Pigment	436.7 ± 1.74^a
Biosurfactant	47.98 ± 2.53^c
Sovaldi	1064.4 ± 1.41^b

All values are expressed as mean±SE. Different letters are significantly different within the same column at p<0.05.

- *In vitro* HCV and HBV infection and treatment

The viral host cells, human PBMCs ($1x10^6$ cells) except negative control wells, were incubated with infected serum with either HCV or HBV, then the untreated and treated infected cells were quantitative analyzed for the intracellular HCV and HBV as shown in **Table (16).**

Table 16: Quantitative anti-HCV and anti-HBV activities of pigment and biosurfactant by the fully automated Cobas Ampliprep TaqMan Analyzer

Samples	Viral load (IU/ml)	
	HCV	HBV
Positive control	$5.09x10^4$	$5.78x10^4$
Pigment	Negative	$0.5x10^4$
Biosurfactant	$1.94x10^4$	$5.75x10^4$
Sovaldi	$1.85x10^3$	$6.95x10^4$

- *In vitro* **inhibition of albumin denaturation assay**

Different concentrations of the pigment extract and partially purified biosurfactant solution were tested for their ability to produce anti-inflammatory agent(s) using a preliminary screening test "inhibition of albumin denaturation assay". The different concentrations of both pigment (5, 10, 15, 20, 25, 30, and 1000 µg/ml) and biosurfactant (1, 2, 3, 4, 5, 6, and 1000 µg/ml) solutions were prepared then tested for their anti-inflammatory activity (the concentrations were prepared according to the safe dose of both pigment and biosurfactant on normal human cells). The results presented in **Table (17)** showed that only the high concentration of the pigment (1000 µg/ml), which is off the range of the safe dose, was able to protect the tested protein (bovine serum albumin) from denaturation compared to diclofenac sodium (a standard anti-inflammatory drug) and represented 40% as shown in **Table (18)**.

Effect of different concentrations on growth of pathogenic bacteria and yeast

The antagonistic effect of different concentrations of both pigment and partially purified biosurfactant solutions of *Natrialba* sp. M6 was examined against *Bacillus subtilis* ATCC 6633, *Klebsiella pneumoniae* ATCC 13883, *Pseudomonas aeruginosa* ATCC 9027, *Escherichia coli* NCTC 10418, *Staphylococcus aureus* ATCC 6538, *Vibrio cholera* ATCC 55188, and *Candida albicans* ATCC 10231 using agar well-cut diffusion assay technique under the following growing conditions: 37°C, 24 h incubation period. The concentrations were used within the safe dose of the normal human cells (30 µg/ml pigment and 6 µg/ml biosurfactant) and the results revealed that, there was no reduction in the survivors as the concentration increased within the safe dose as shown in **Table (18)**.

- **Anti-fungal assay of the sterile pigment and partially purified biosurfactant solutions**

The sterile pigment and biosurfactant extracts were examined for their antifungal activity. Under our experimental condition 37°C and 24 h incubation period, the examined extracts did not show any antifungal activity against *Aspergillus flavus* ATCC 77708, *Rhizoctonia solani* ATCC 10532 and *Fusarium solani* ATCC 11233 at the highest safe doses (30 µg/ml pigment and 6 µg/ml biosurfactant) as shown in **Table (19)**.

Table 17: The screening process for anti-inflammatory activity of the test compound using inhibition of albumin denaturation assay.

The test compound	Concentration µg/ml	Absorption at 550nm Wavelength	Inhibition of protein denaturation (%)
Pigment extract solution	5	0.982	ND
	10	0.935	ND
	15	0.845	ND
	20	0.834	ND
	25	0.831	ND
	30	0.829	ND
	1000	0.492	40±12
Biosurfactant extract solution	1	0.999	ND
	2	1.283	ND
	3	1.115	ND
	4	0.983	ND
	5	0.891	ND
	6	0.899	ND
	1000	1.233	ND
Diclofenac sodium	1000	0.075	90±9

- Bovine serum albumin without any treatment was used as a control.

Table 18: Anti-bacterial activity of the pigments and biosurfactants extracts against different pathogenic bacteria

Bacterial pathogens	Pigment conc. µg/ml						Biosurfactant conc. µg/ml					
	5	10	15	20	25	30	1	2	3	4	5	6
Bacillus subtilis	-ve	-ve	-ve	-ve	-ve	-ve	-ve	-ve	-ve	-ve	-ve	-ve
Klebsiella pneumoniae	-ve	-ve	-ve	-ve	-ve	-ve	-ve	-ve	-ve	-ve	-ve	-ve
Pseudomonas aeruginosa	-ve	-ve	-ve	-ve	-ve	-ve	-ve	-ve	-ve	-ve	-ve	-ve
Escherichia coli	-ve	-ve	-ve	-ve	-ve	-ve	-ve	-ve	-ve	-ve	-ve	-ve
Staphylococcus aureus	-ve	-ve	-ve	-ve	-ve	-ve	-ve	-ve	-ve	-ve	-ve	-ve
Vibrio cholera	-ve	-ve	-ve	-ve	-ve	-ve	-ve	-ve	-ve	-ve	-ve	-ve
Candida albicans	-ve	-ve	-ve	-ve	-ve	-ve	-ve	-ve	-ve	-ve	-ve	-ve

Table 19: Anti-fungal activity of the pigments and biosurfactants extracts against different pathogenic fungi

Fungal pathogens	Pigments extract 30 µg/ml	Biosurfactants extract 6 µg/ml
Aspergillus flavus	-ve	-ve

Rhizoctonia solani	-ve	-ve
Fusarium solani	-ve	-ve

Anti-oxidant capacity of pigment and partially purified biosurfactant solutions

The antioxidant capacity of the pigment and the biosurfactant solutions is shown in **Table (20)**. As clearly observed from the results of the present investigation, the pigment exhibited antioxidant capacity at 1000 µg/ml concentration, which is equal to 990000 µmol/g. This value was higher than that of the biosurfactant, which gave an antioxidant capacity equal to 30000 µmol/g at the same concentration using the phosphomolybdenum reagent.

Table 20: Antioxidant capacity µmol/g equivalent of pigment and biosurfactant.

Antioxidant activity	µmol/g equivalent BHT standard
Pigment extract solution (1000 µg/ml)	990000
Biosurfactant extract solution (1000µg/ml)	30000

Anti-biofouling activity of *Natrialba* sp. M6 partially purified biosurfactant and pigment

In this experiment, the antifouling activity of the pigment and biosurfactant was tested. The result obtained after microscopic examination of the dried cover glasses showed that both products inhibited the biofouling formation. Microscopic examination of the inhibited cover glasses showed that, lower counts of bacterial biofilms on the treated panels (treated glass slides), when compared with untreated ones. By increasing the concentration of the biosurfactant and pigment from 500 µg to 1000 µg, the number of adhered bacteria decreased. i.e. the pigment and the partially purified biosurfactant have a remarkable valuable effect on biofouling reduction **(Fig. 13)**.

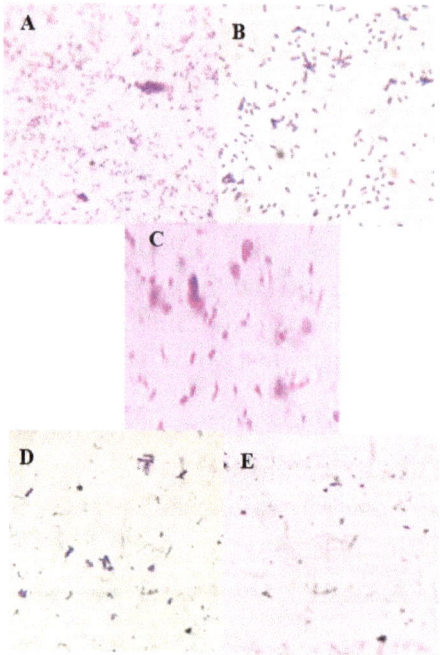

Fig. 13: Effect of the biosurfactant and pigment extract on biofilm formation (A) in the absence of the extract, (B) in the presence of 500 µg/ml of the biosurfactant, (C) in the presence of 1000 µg/ml of the biosurfactant, (D) in the presence of 500 µg/ml of the pigment, (E) in the presence of 1000 µg/ml of the pigment.

Discussion

In the present study, *Natrialba* sp. M6 strain was able to produce biosurfactant and its production was optimized. Strain M6 produced the maximum biosurfactant concentration in a basal medium supplemented with 200 g/l NaCl, adjusted to pH 11, and grown at 37°C. Meanwhile, pigmentation was increased in a medium supplemented with 150g/l NaCl, pH 10 and incubated at 45°C. It was previously reported that

Natrialba sp. C21 produced biosurfactant in a medium containing 250 g/l, pH 8, and grown at 40°C **(Khemili-Talbi et al., 2015).** This variation in growth conditions could be attributed to the alkaliphilic nature of the M6 strain under the current investigation. These results indicate that the biosurfactant and pigment are produced via independent biosynthetic pathways, and pH, temperature, and NaCl concentration are all significant factors in the optimization of their production.

Sequential optimization approaches were employed to maximize the biosurfactant production from *Natrialba* sp. M6. Different variables that assumed to affect the biosurfactant production were investigated using Plackett-Burman Design **(Plackett & Burman, 1946).** This method was previously employed for the biosurfactant production by a marine *Bacillus* sp. **(Mabrouk et al., 2014),** and *Natrialba* sp. **(Khemili-Talbi et al., 2015).** The selected variables included several medium components such as; carbon source, nitrogen source, and minerals, as well as physical conditions. Sixteen experimental trials were conducted and the biosurfactant production was determined by measuring the surface tension to calculate the response as described by **(Liu et al., 2015).** In the current study, the effect of using different carbon sources in the culture medium (glucose and glycerol) on the production of biosurfactant was investigated. Based on the main effect calculated from the regression coefficients for the 14 variables, it was found that the significant factors affecting the biosurfactant production by *Natrialba* sp. M6 were agitation, ammonium nitrate, sodium chloride, glucose and yeast extract.

The type of carbon source has been reported in literature as a vital limiting factor in microbial biosurfactant production process **(Pal et al., 2009).** Several studies had investigated the effect of using different carbon sources on the biosurfactant production efficiency. For instant, the use of different hydrophilic carbon sources (glucose and glycerol) by *Pseudomonas aeruginosa* in the production of rhamnolipid (biosurfactant) was investigated, glycerol was found to be the best carbon source for its production (**Mehdi et al., 2011**). Meanwhile, (**Mouafo et al., 2018**) tested another two carbon sources including sugar-cane molasses and glycerol as substrates for the production of biosurfactants by *Lactobacillus* spp. They also concluded that glycerol was the favorable carbon source for the production of glycolipids biosurfactant. Sugar cane molasses supported the production of a biosurfactant by a marine *Bacillus* sp. (**Mabrouk et al., 2014**). The potential for surfactant production was investigated in several studies. For example, the extreme halophilic archaeon *Haloferax* sp. MSNC14 was able to produce biosurfactants in the presence of individual hydrocarbon substrates (**Djeridi et al., 2013**). In addition, (**Abbasi et al., 2013**) determined the optimum conditions for maximum production of microbial biosurfactant by *P. aeruginosa* and the results of their study showed that soybean oil and $NaNO_3$ were the most effective carbon and nitrogen sources, respectively. However, yeast extract (used as a complementary nitrogen source) was vital for maximum biosurfactant production. (**Kumar et al., 2015**). reported that pH, and glycerol has profound effect on the biosurfactant production by *Pseudomonas* sp. In the contrast to these studies, we found that glucose and ammonium nitrate were more effective than pH and glycerol, but both glycerol and pH were selected for further optimization step, since both ammonium nitrate and glucose showed negative effect on the biosurfactant production with a low value (-1), which was equal to zero. Also high glucose concentration was found inhibited the production of biosurfactant due to the formation of acidic metabolites (**Heryani & Putra, 2017**). Also (**Mehdi et al., 2011**) found that biosurfactant production by *Pseudomonas aeruginosa* was limited by high concentration of glucose, nitrogen, iron, maganesium, phosphorus, sulfur, and calcium. This was attributed to the promoting effect of glucose on the archaeal growth in the exponential phase leading to shortened lag phase and increased cell concentration. The increase in the cell growth resulted in biosurfactant over production **(Roy, 2017)**.

Among the physicochemical factors, only temperature and shaking had a significant effect on biosurfactant production **(Abbasi et al., 2013)**.

On the model level, the correlation measures for estimating the regression equation are the multiple correlation coefficient, R, and the determination coefficient, R^2. The closer the value of R to 1, the stronger is the correlation between the experimental and the predicted values. In the present experiment, the value of R was 0.96 indicating a high correlation between the experimental and predicted results, likewise, R^2 for the two-level-screening experiment of biosurfactant production was 0.92.

Box-Behnken design was applied, in the current study, to evaluate the effect of pH, glycerol, agitation, and sodium chloride as independent variables on the biosurfactant production and to identify the optimum response region for these factors as described by **(Box & Behnken, 1960)**.

For predicting the optimum point, a second order polynomial function was fitted to correlate the relationship between variables and responses. The obtained polynomial model and the correlated three-dimensional graphs showed that the effects of the four significant variables; pH, glycerol, agitation, and NaCl is due to the significant interaction between them. The optimal levels of the four factors were also estimated by solving the polynomial model, and were found to be the best production at pH 12, 3% glycerol, 150 rpm agitation, and 20.8% NaCl, by which Appling the above mentioned parameters gave the highest biosurfactant concentration. In previous study, **(Kumar et al., 2015)** reported that an optimum combination of variables (pH: 7.3, sawdust: 7.656, 1.5% glycerol) was achieved for the production of biosurfactant by *Pseudomonas putida* MTCC 2467 through predicted plot of full quadratic model.

Carotenoids are members of a group of naturally occurring pigments that have received considerable attention due to their biotechnological applications and their potential beneficial effects on human health **(Pathak & Sardar, 2012)**. Halophilic archaea, inhabiting salty environments such as marshes or salty ponds, from where NaCl is obtained for human consumption, possess high carotenoids production **(Oren, 2013, Gupta et al., 2015)**. Indeed, the use of haloarchaea for the commercial production of carotenoids is of industrial advantage compared to other microorganisms as they tolerate harsh manufacturing processes **(Rodrigo-Baños et al., 2015)**.

The results presented in the current investigation revealed that NaCl concentration was the most important factor that affected cell growth and total carotenoids production by *Natrialba* sp. M6. It was previously reported that NaCl concentration is the most important factor in determining haloalkaliphilic **(Hamidi et al., 2014)**. In addition, biomass and total carotenoids production by the M6 strain were at their optimum levels, when 15 to 25% NaCl (w/v) was added to the production media (pH 10) while cultures were incubated at 45°C. **(Calegari-Santos et al., 2016)** reported that *Haloferax mediterranei* ATCC 33500 produced about 20 times more pigments in liquid medium containing 15% NaCl than the medium with 25% NaCl, incubated at 45°C, and pH range 7-9.

The growth medium pH was another important factor affecting the biomass and total carotenoid production by *Natrialba* sp. M6 strain. It has been reported that pH significantly influences cell growth and total carotenoid production in many microorganisms **(Oren, 2012)**. Most of the extremely halophilic archaea grow in neutral media, however a small group of extremely halophilic archaea has shown an optimum growth in alkaline media (pH 8.5 and greater), and therefore was considered as haloalkaliphiles **(Jehlička et al., 2013)**.

In our work maximal carotenoid production was attained by using old cultures of *Natrialba* sp. M6 and adjusting pH between 9 –11. It is noteworthy that old cultures showed significant increase in the production of carotenoid **(Torregrosa-Crespo et al., 2018)**.

Hence, NaCl, pH and culture age were further investigated as significant independent variables in Box-Behnken design to identify the optimum response region for carotenoid biosynthesis in *Natrialba* sp. M6 strain. Then, a second order polynomial function was fitted to correlate the relationship between the variables and their responses. The results indicated that the optimal levels of the three factors were 25% NaCl, pH 10.07, and 9 days culture age. **(Hamidi et al., 2014)** used Box-Behnken design to determine the optimum factors affecting carotenoids production by *Halorubrum* sp. TBZ126, they reported that the estimated coefficients and the corresponding *p*-values indicate that X1 (Temperature), X2 (pH) and X3 (salinity) were significant factors for both cell growth and carotenoid production by this strain. The authors found that the optimum conditions for carotenoid production were 32°C, pH 7.94, and 20.55% NaCl.

In our study, the partially purified biosurfactant extract was characterized using GC–MS analysis as recommended by **(Jerković et al., 2015).** The presented data revealed that the recovered biosurfactant extracts contained a diverse group consists of 34 compounds, mostly biosurfactant as they are showing bipolar nature, hydrophobic hydrocarbon chain and charged function group also most of these compounds are fatty acids. In addition, total proteins, carbohydrates, and lipids analyses of the recovered biosurfactant extract were determined. The results indicated that protein and lipid contents were higher than carbohydrate, which indicate that the biosurfactant was mostly lipoprotein or lipopeptide biosurfactant. **(Rufino et al., 2014)** reported that preliminary chemical characterization of the biosurfactant produced from *Candida lipolytica* UCP 0988 revealed that the examined agent produced a lipoprotein material, which consisted of protein (50%), lipid (20%), and carbohydrates (8%). Meanwhile, *Candida sphareica* produced glycolipid product that consists of 70% lipids and 15% carbohydrates.

Furthermore, in the current study, the orange-pigment extract was characterized using both Raman Spectroscopy and FTIR as chemical typing methods, the analysis of the resultant peaks was attributed to the carotenoids group due to the presence of methyl group (CH_3) and conjugated double bond.

The partially purified extracted biosurfactant and pigment were then used to explore their potential industrial and medicinal applications, as an additional aim of this study. The anticancer activity of the partially purified biosurfactant and pigment extracts of *Natrialba* sp. M6 strain was investigated. It was found that the pigment had a greater anticancer activity than both the M6 biosurfactant, and the standard chemotherapy (5-fluorouracil), when it was tested against four different cancer cell lines (Caco-2, MCF-7, HepG-2, Hela). These results were attributed to the high IC_{50} and EC_{100} values, which are indicators for how safe are the pigments toward normal cells. In comparison, Biosurfactant and currently used chemotherapy gave lower IC_{50} and EC_{100} values, Similarly **(Janek et al., 2013)** investigated induction of lipopeptide biosurfactant pseudofactin II produced by some bacteria apoptosis of Melonoma (M) A 375 cells by specific interaction with the plasma membrane of infected M cells. Also the biological roles of the carotenoids produced by haloarchaea in haloarchaeal were studied by **(Sikkandar et al., 2013),** and they reported that *Halobacterium salinarum* produces various pigments such as phytoene, β-carotene, lycopene and derivatives of bacterioruberin and salinixanthin. These pigments were tested for their cytotoxic properties against human liver cancer cell lines and showed dose-dependent increase in cytotoxicity of the carotenoids on these cells, suggesting probable anti-cancer properties. More studies are required to explore the potential medicinal effects of the archaeal carotenoids and biosurfactants on cancer cells.

Therefore, cancer selectivity index was studied for both M6 pigment and biosurfactant against four cancer cell lines (Caco-2, MCF-7, HepG-2, Hela), and was compared to the results of standard chemotherapy. The results showed high selectivity index values for the tested pigment toward the tested cancer cell lines, whereas, the selectivity index of the archaeal biosurfactant showed lower values than the chemotherapy selectivity index **(Hegazy et al., 2022)**. These results indicate that the archaea pigment has a higher selectivity towards cancer cells, compared to the currently used chemotherapy (FU), and suggesting potential medicinal applications of the archaeal carotenoids against cancer. In previous studies, the potential anticancer properties of bacterial carotenoids were demonstrated. For example, the bacterial carotenoids, which were isolated from two strains of halophilic bacteria- *Halobacterium salinarium* and *Haloferox volcanii*, were proven to have antioxidant activity and anticancer properties against human liver cancer cell lines **(Sikkandar et al., 2013)**. In addition, **(Maithili et al., 2013)** studied the anticancer properties of the red pigment extracted from *Serritia marcenscens* against two different types of cancer cell lines (breast and prostate cancers), and they showed that the pigment had stronger anticancer properties against prostate cancer cell lines.

Furthermore, to identify the mechanism of cell death, the annexin stain was used in the present study to investigate the effect of carotenoid on the cancer cell lines. The high percentage of annexin-stained cancer cells treated with pigments was an indication for the pigment-mediated apoptosis cell death in different treated cancer cells **(Hegazy et al., 2020)**. In a similar study, **(Li et al., 2018)** investigated the apoptosis effect of prodigiosin pigment, which is produced by *Serratia marcescens*, using the annexin staining technique. Their results showed the ability of prodigiosin of inducing apoptosis in cancer cells similar to prodigiosin standard. This result strongly supports the obtained results in our study.

Furthermore, the antiviral activity of both biosurfactant and pigment isolated from *Natrialba* sp. M6 was also studied against both hepatitis HCV and HBV activities. The results indicated that the pigment was completely efficient in inhibiting the activity of virus C compared to the biosurfactant treatment that decreased the count of the virus from 5.09×10^4 to 1.94×10^4. On the other hand, HBV was activated when cells were treated with both M6 biosurfctant and Sovaldi, but, surprisingly, the viral activity was decreased with the pigment treatment from 5.78×10^4 to 0.5×10^4 **(Hegazy et al., 2020)**, **(Zhou et al., 2016)** indicated that prodigiosin pigment produced *Serratia marcescens* possesses significant antiviral activity against Bombyx mori nucleopolyhedrovirus (BmNPV)- infected cells *in vitro* under prodigiosin treatment, both progeny virus production and viral DNA replication were significantly inhibited.

The antimicrobial activity against pathogenic bacteria and fungi was studied and the presented results showed that there is no detected effect of both biosurfactant and pigment treatments within the safe dose. The lack of antimicrobial activity of these compounds at low concentrations might be because these compounds have antioxidant activity, and when it is applied within the save dose levels, a cell stimulating effect of the living cells may occur **(Hou & Cui, 2018)**. However, the effect of these compounds at higher concentrations might depict higher microbial toxicities, since the pigment has showed cytotoxic effect for both cancer cells lines and HCV. However, further antimicrobial studies are required to test this hypothesis. The antimicrobial effects of carotenoids were studied elsewhere, and the results indicated that pigment produced by *Halolactobacillus alkaliphilus* MSRD1 was highly inhibited *Staphlyococcus aureus* and *Salmonella typhi* **(Suresh et al., 2015)**. Also **(Ramesh et al., 2017)** investigated the anti-bacterial activity of the pigment of many bacterial strains including (*Bacillus flexus, Micrococcus luteus, Photobacterium ganghwense, Stenotrophomonas maltpphilia and Vibrio* sp) against many human pathogenic bacteria.

Anti-inflammatory refers to the property of a substance or treatment that reduces inflammation. Over the last years, the marine environment introduced compounds with anti-inflammatory activities. The need for novel anti-inflammatory drugs is increasing due to issues concerning the serious side effects of the traditional anti-inflammatory drugs **(D'Orazio et al., 2012).** Hence, in the present study, the protein denaturation bioassay was adopted for the assessment of the *in vitro* anti-inflammatory properties of the *Natrialba* sp. M6 biosurfactant and pigment. Denaturation of tissue proteins is one of the most common cause of inflammatory reactions, and the production of the auto antigens in certain arthritic diseases may be due to the denaturation of proteins *in vivo* **(Umapathy et al., 2010).**

The results obtained in the current work indicated that the pigment showed anti-inflammatory effect and prevented the denaturation of BSA protein at a concentration much higher than the safe dose, which indicates that the pigments interacting with the aliphatic regions around the lysine residue on the BSA. This could be interesting anti-inflammatory with anticancer activity such as the polyphenols, phenyl propanoids and the disulphides **(Williams et al., 2008).** In addition, **(Heo et al., 2010)** suggested the usage of fucoxanthin (axanthophyll), which is one of two major divisons of the carotenoid group, as a useful therapeutic approach for various inflammatory diseases.

There is some evidence supporting that carotenoids from haloarchaea are as efficient as those antioxidant compounds produced by other microorganisms **(Sikkandar et al., 2013).** The pigment and biosurfactant that is produced by *Natrialba* sp. M6 in the current study were also tested for their free radical scavenging activity. The results confirmed the previously documented antioxidant activity of secondary metabolites belonging to the carotenoids and biosurfactants, which was extracted from archaea as an alternative source of antioxidant carotenoids **(Hou & Cui, 2018).**

Furthermore, it was found in this study that the more the concentration of both biosurfactant and carotenoid (500, 1000µg/ml) the more inhibition effect of bacterial biofilm formation exists, the extract already reduced the number of bacterial cells acting as anti-biofouling agent. This anti-biofouling effect was achieved when the pigment and the biosurfactant was used in concentration much higher than the concentration tested within the safe dose. This could be indication for the antimicrobial effect of the pigment and biosurfactant produced by *Natrialba* sp M6, however further studies are required to elucidate these effects. In addition, **(Jemil et al., 2017)** suggested the inhibition of biofilm formation by lipopeptides, which are produced by *Bacillus methylotrophicus* DCS1. Meanwhile, **(Rodrigo-Baños et al., 2015)** suggested the antifouling activity of carotenoids extracted from haloarchaea and the ability to inhibit biofilm formation.

Conclusion

The archaeon isolate *Natrialba* sp. M6 was tested for its ability to produce biosurfactants and pigments. The FTIR and Raman investigations illustrated the presence of aliphatic compounds and organic carotenoids in the pigment extract, respectively. The extracted pigment showed a significant cytotoxicity, anti-inflammatory effects, and anti-biofouling activity in addition to the ability to seduce apoptosis in cancer cells.

References

Abbasi H, Sharafi H, Alidost L, Bodagh A, Zahiri HS, Noghabi KA, 2013. Response surface optimization of biosurfactant produced by Pseudomonas aeruginosa MA01 isolated from spoiled apples. *Preparative Biochemistry and Biotechnology* **43**, 398-414.

Abdel-Fattah Y, El-Enshasy H, Soliman N, El-Gendi H, 2009. Bioprocess development for production of alkaline protease by Bacillus pseudofirmus Mn6 through statistical experimental designs. *J Microbiol Biotechnol* **19**, 378-86.

Abdel-Fattah YR, El Enshasy H, Anwar M, Omar H, Abolmagd E, Zahra R, 2007. Application of factorial experimental designs for optimization of cyclosporin a production by Tolypocladium inflatum in submerged culture. *Journal of microbiology and biotechnology* **17**, 1930.

Bavya M, Mohanapriya P, Pazhanimurugan R, Balagurunathan R, 2011. Potential bioactive compound from marine actinomycetes against biofouling bacteria.

Box GE, Behnken DW, 1960. Some new three level designs for the study of quantitative variables. *Technometrics* **2**, 455-75.

Calegari-Santos R, Diogo RA, Fontana JD, Bonfim TMB, 2016. Carotenoid production by halophilic archaea under different culture conditions. *Current microbiology* **72**, 641-51.

D'orazio N, Gammone MA, Gemello E, De Girolamo M, Cusenza S, Riccioni G, 2012. Marine bioactives: Pharmacological properties and potential applications against inflammatory diseases. *Marine drugs* **10**, 812-33.

Djeridi I, Militon C, Grossi V, Cuny P, 2013. Evidence for surfactant production by the haloarchaeon Haloferax sp. MSNC14 in hydrocarbon-containing media. *Extremophiles* **17**, 669-75.

Gupta RS, Naushad S, Baker S, 2015. Phylogenomic analyses and molecular signatures for the class Halobacteria and its two major clades: a proposal for division of the class Halobacteria into an emended order Halobacteriales and two new orders, Haloferacales ord. nov. and Natrialbales ord. nov., containing the novel families Haloferacaceae fam. nov. and Natrialbaceae fam. nov. *International journal of systematic and evolutionary microbiology* **65**, 1050-69.

Haba E, Espuny M, Busquets M, Manresa A, 2000. Screening and production of rhamnolipids by Pseudomonas aeruginosa 47T2 NCIB 40044 from waste frying oils. *Journal of Applied microbiology* **88**, 379-87.

Hamidi M, Abdin MZ, Nazemyieh H, Hejazi MA, Hejazi MS, 2014. Optimization of total carotenoid production by Halorubrum sp. TBZ126 using response surface methodology. *J. Microb. Biochem. Technol* **6**, 286-94.

Hasan M, Das R, Khan A, Hossain M, Rahman M, 2009. The determination of antibacterial and antifungal activities of Polygonum hydropiper (L.) root extract. *Advances in Biological Research* **3**, 53-6.

Hegazy GE, Abu-Serie MM, Abo-Elela GM, *et al.*, 2020. In vitro dual (anticancer and antiviral) activity of the carotenoids produced by haloalkaliphilic archaeon Natrialba sp. M6. *Scientific reports* **10**, 5986.

Hegazy GE, Abu-Serie MM, Abou-Elela G, *et al.*, 2022. Bioprocess development for biosurfactant production by Natrialba sp. M6 with effective direct virucidal and anti-replicative potential against HCV and HSV. *Scientific reports* **12**, 16577.

Heo S-J, Yoon W-J, Kim K-N, *et al.*, 2010. Evaluation of anti-inflammatory effect of fucoxanthin isolated from brown algae in lipopolysaccharide-stimulated RAW 264.7 macrophages. *Food and chemical toxicology* **48**, 2045-51.

Heryani H, Putra MD, 2017. Kinetic study and modeling of biosurfactant production using Bacillus sp. *Electronic Journal of Biotechnology* **27**, 49-54.

Hou J, Cui H-L, 2018. In vitro antioxidant, antihemolytic, and anticancer activity of the carotenoids from halophilic archaea. *Current microbiology* **75**, 266-71.

Iqbal S, Khalid Z, Malik K, 1995. Enhanced biodegradation and emulsification of crude oil and hyperproduction of biosurfactants by a gamma ray-induced mutant of Pseudomonas aeruginosa. *Letters in Applied Microbiology* **21**, 176-9.

Janek T, Krasowska A, Radwańska A, Łukaszewicz M, 2013. Lipopeptide biosurfactant pseudofactin II induced apoptosis of melanoma A 375 cells by specific interaction with the plasma membrane. *PloS one* **8**, e57991.

Jehlička J, Edwards H, Oren A, 2013. Bacterioruberin and salinixanthin carotenoids of extremely halophilic Archaea and Bacteria: a Raman spectroscopic study. *Spectrochimica Acta Part A: Molecular and Biomolecular Spectroscopy* **106**, 99-103.

Jemil N, Ben Ayed H, Manresa A, Nasri M, Hmidet N, 2017. Antioxidant properties, antimicrobial and anti-adhesive activities of DCS1 lipopeptides from Bacillus methylotrophicus DCS1. *BMC microbiology* **17**, 1-11.

Jerković I, Tuberoso CIG, Baranović G, *et al.*, 2015. Characterization of summer savory (Satureja hortensis L.) honey by physico-chemical parameters and chromatographic/spectroscopic techniques (GC-FID/MS, HPLC-DAD, UV/VIS and FTIR-ATR). *Croatica Chemica Acta* **88**, 15-22.

Khemili-Talbi S, Kebbouche-Gana S, Akmoussi-Toumi S, Angar Y, Gana ML, 2015. Isolation of an extremely halophilic arhaeon Natrialba sp. C21 able to degrade aromatic compounds and to produce stable biosurfactant at high salinity. *Extremophiles* **19**, 1109-20.

Kumar AP, Janardhan A, Radha S, Viswanath B, Narasimha G, 2015. Statistical approach to optimize production of biosurfactant by Pseudomonas aeruginosa 2297. *3 Biotech* **5**, 71-9.

Kumar T, Aparna H, 2014. Anti-biofouling activity of Prodigiosin, a pigment extracted from Serratia marcescens. *Int. J. Curr. Microbiol. App. Sci* **3**, 712-25.

Li D, Liu J, Wang X, *et al.*, 2018. Biological potential and mechanism of prodigiosin from Serratia marcescens subsp. lawsoniana in human choriocarcinoma and prostate cancer cell lines. *International Journal of Molecular Sciences* **19**, 3465.

Liu J-F, Mbadinga SM, Yang S-Z, Gu J-D, Mu B-Z, 2015. Chemical structure, property and potential applications of biosurfactants produced by Bacillus subtilis in petroleum recovery and spill mitigation. *International Journal of Molecular Sciences* **16**, 4814-37.

Louis KS, Siegel AC, 2011. Cell viability analysis using trypan blue: manual and automated methods. *Mammalian cell viability: methods and protocols*, 7-12.

Mabrouk ME, Youssif EM, Sabry SA, 2014. Biosurfactant production by a newly isolated soft coral-associated marine

Bacillus sp. E34: Statistical optimization and characterization. *Life Sci. J* **11**, 756-68.

Maithili A, Samir S, Gayatri M, Smera S, Sangeeta S, 2013. Comparitive In-vitro cytotoxicity of red pigment extract of Serratia marcescens on breast and prostate cancer cell lines. *International Journal of Current Pharmaceutical Research* **5**, 140-3.

Mehdi S, Dondapati JS, Rahman PK, 2011. Influence of nitrogen and phosphorus on rhamnolipid biosurfactant production by Pseudomonas aeruginosa DS10-129 using glycerol as carbon source. *Biotechnology* **10**, 183-9.

Mizushima Y, Kobayashi M, 1968. Interaction of anti-inflammatory drugs with serum proteins, especially with some biologically active proteins. *Journal of Pharmacy and Pharmacology* **20**, 169-73.

Morikawa M, Hirata Y, Imanaka T, 2000. A study on the structure–function relationship of lipopeptide biosurfactants. *Biochimica et Biophysica Acta (BBA)-Molecular and Cell Biology of Lipids* **1488**, 211-8.

Mosmann T, 1983. Rapid colorimetric assay for cellular growth and survival: application to proliferation and cytotoxicity assays. *Journal of immunological methods* **65**, 55-63.

Mouafo TH, Mbawala A, Ndjouenkeu R, 2018. Effect of different carbon sources on biosurfactants' production by three strains of Lactobacillus spp. *BioMed research international* **2018**.

Oren A, 2012. Taxonomy of the family Halobacteriaceae: a paradigm for changing concepts in prokaryote systematics. *International journal of systematic and evolutionary microbiology* **62**, 263-71.

Oren A, 2013. Salinibacter: an extremely halophilic bacterium with archaeal properties. *FEMS microbiology letters* **342**, 1-9.

Pal MP, Vaidya BK, Desai KM, Joshi RM, Nene SN, Kulkarni BD, 2009. Media optimization for biosurfactant production by Rhodococcus erythropolis MTCC 2794: artificial intelligence versus a statistical approach. *Journal of Industrial Microbiology and Biotechnology* **36**, 747-56.

Pathak AP, Sardar AG, 2012. Isolation and characterization of carotenoid producing Haloarchaea from solar saltern of Mulund, Mumbai, India.

Plackett RL, Burman JP, 1946. The design of optimum multifactorial experiments. *Biometrika* **33**, 305-25.

Ramesh C, Mohanraju R, Murthy K, Karthick P, 2017. Molecular characterization of marine pigmented bacteria showing antibacterial activity.

Rodrigo-Baños M, Garbayo I, Vílchez C, Bonete MJ, Martínez-Espinosa RM, 2015. Carotenoids from Haloarchaea and their potential in biotechnology. *Marine drugs* **13**, 5508-32.

Roy A, 2017. Effect of various culture parameters on the bio-surfactant production from bacterial isolates. *Journal of Petroleum & Environmental Biotechnology* **8**, 350.

Rufino RD, De Luna JM, De Campos Takaki GM, Sarubbo LA, 2014. Characterization and properties of the biosurfactant produced by Candida lipolytica UCP 0988. *Electronic Journal of Biotechnology* **17**, 34-8.

Sakat S, Juvekar AR, Gambhire MN, 2010. In vitro antioxidant and anti-inflammatory activity of methanol extract of Oxalis corniculata Linn. *Int J Pharm Pharm Sci* **2**, 146-55.

Sikkandar S, Murugan K, Al-Sohaibani S, Rayappan F, Nair[o] A, 2013. Halophilic Bacteria-A Potent Source of Carotenoids. *Journal of Pure and Applied Microbiology* **7**, 2825-30.

Suresh M, Renugadevi B, Brammavidhya S, Iyapparaj P, Anantharaman P, 2015. Antibacterial activity of red pigment produced by Halolactibacillus alkaliphilus MSRD1—an isolate from seaweed. *Applied biochemistry and biotechnology* **176**, 185-95.

Thaniyavarn J, Chongchin A, Wanitsuksombut N, *et al.*, 2006. Biosurfactant production by Pseudomonas aeruginosa A41 using palm oil as carbon source. *The Journal of General and Applied Microbiology* **52**, 215-22.

Thrane J-E, Kyle M, Striebel M, *et al.*, 2015. Spectrophotometric analysis of pigments: a critical assessment of a high-throughput method for analysis of algal pigment mixtures by spectral deconvolution. *PloS one* **10**, e0137645.

Torregrosa-Crespo J, Montero Z, Fuentes JL, *et al.*, 2018. Exploring the valuable carotenoids for the large-scale production by marine microorganisms. *Marine drugs* **16**, 203.

Umapathy E, Ndebia E, Meeme A, *et al.*, 2010. An experimental evaluation of Albuca setosa aqueous extract on membrane stabilization, protein denaturation and white blood cell migration during acute inflammation. *J Med Plants Res* **4**, 789-95.

Valgas C, Souza SMD, Smânia EF, Smânia Jr A, 2007. Screening methods to determine antibacterial activity of natural products. *Brazilian journal of microbiology* **38**, 369-80.

Williams L, O'connar A, Latore L, *et al.*, 2008. The in vitro anti-denaturation effects induced by natural products and non-steroidal compounds in heat treated (immunogenic) bovine serum albumin is proposed as a screening assay for the detection of anti-inflammatory compounds, without the use of animals, in the early stages of the drug discovery process. *West Indian Medical Journal* **57**.

Youssef NH, Duncan KE, Nagle DP, Savage KN, Knapp RM, Mcinerney MJ, 2004. Comparison of methods to detect biosurfactant production by diverse microorganisms. *Journal of microbiological methods* **56**, 339-47.

Zhou W, Zeng C, Liu R, *et al.*, 2016. Antiviral activity and specific modes of action of bacterial prodigiosin against Bombyx mori nucleopolyhedrovirus in vitro. *Applied microbiology and biotechnology* **100**, 3979-88.

Plackett-Burman optimized conditions for electricity generation by Halophilic archaeon *Natrialba* sp. GHMN55: single and stacked microbial fuel cells

Abstract

In this work, the archaeon *Natrialba* sp. GHMN55 has been used as a biocatalyst for the generation of electricity through the microbial fuel cells. Multiple factors have been tested for their effect on the power production after the optimization using Plackett-Burman experimental design. Of such factors; casein, inoculum age, and resistor value showed a positive effect on the electricity production compared with the existence of mediator and pH value that showed negative effects. This optimized design resulted in ten-fold reduction of the incubation time from 20 days to 2 days with the maximum production of 200 mV. Stacked MFCs have been also tested in order to increase the power production, which showed that the series configuration was the preferred one compared with the parallel or single cell configurations **(Hegazy et al., 2022)**.

Keywords: Plackett-Burman design, MFC, Electricity generation, Archaea, *Natrialba* sp.

Introduction

The microbial fuel cells are electrochemical systems that use the microbes to turn the energy stored in the chemical bonds of organic molecules into electrical current with no need for combustion **(Bishoge et al., 2019)**. These microbes are mostly working under anaerobic conditions in the anodic chamber, where the protons pass through the bridge membrane into the aerobic cathode chamber to form water at the presence of oxygen molecules, while the electrons are passing through the external circuit generating the electrical current **(Bond & Lovley, 2003) (Ulusoy & Dimoglo, 2018, Xu et al., 2019, Cao et al., 2019)**.

A lot of factors such as the microorganism, the electrodes reactions, the reactor design, the internal resistance, and the organic matter are affecting the performance of MFCs **(Kaur et al., 2014)**. Moreover, any external physical or chemical changes in addition to the growth conditions and the microorganisms can lead to alterations in the physiological parameters that leading to inhibiting the metabolism and the growth of the microorganisms with subsequent death **(Li et al., 2014)**.

The archaeon *Natrialba* sp. GHMN55 is an extremophilic halophilic archaea that belongs to the family Halobacteriaceae. The members of this family are able to live in extreme habitats such as hypersaline environments that can reach to 5M salt concentration with pH up to 12 **(Hegazy et al., 2020)**.

This study is aiming to investigate the power generation by the halophilic archaeon *Natrialba* sp. GHMN55 **(accession No MW794195)** when used as a biocatalyst in the anodic chamber in a microbial fuel cell using high salt concentration. Moreover, the power production under optimized conditions using Plackett-Burman statistical design was also investigated through the testing of 10 different factors such as NaCl concentration, pH, casein concentration, oxygen incorporation, existence/absence of mediator, existence/absence of magnetic bar, resistor value, inoculum age, inoculum size, and incubation time.

Methods

Microbial isolation (Source and culture medium)
The characterization of El-Hamra lake at Wadi El-Natrun, in addition to the media composition and isolation conditions are the same like the data represented in chapter I **(Hegazy et al., 2020)**.

Preparation of inoculums
Two pre-cultures of *Natrialba* sp. GHMN55 isolate (3 days 'OD$_{600}$ 0.4' and 6 days 'OD$_{600}$ 0.9') were prepared. Both pre-cultures were used as the inoculum with percentage of 10% v/v according to need.

Scanning electron Microscopy (SEM)
The shape of the cells of the archaeon *Natrialba* sp. was shown using JSM 5300 scanning electron microscope (JEOL, USA) at 20 kV after coating with a thin gold film using sputtering device 54 (JFC-1100 E, JEOL, USA) for 12 min.

Architectures of the microbial fuel cells

(a) Single unit MFC

Cheap and waste materials were used for the design of single cell MFC. Two plastic containers with a total volume capacity of 300 ml were used as the anode and cathode chambers of the cell. Both chambers were connected by a bridge of a glass tube filled with agar. Two holes of 0.2 and 1.0 cm diameters were made in each chamber lid in order to allow the passage of the stainless steel wire connections and the agar bridge. After the addition of the bacterial inoculum (10% *Natrialba* sp. GHMN55) and the mediator (phenolphthalein), the solution of the anode chamber was covered with mineral oil in order to inhibit the oxygen penetration, while the saline in the cathode chamber was bubbled continuously with air in order to increase the oxygen supply (**Fig. 1**).

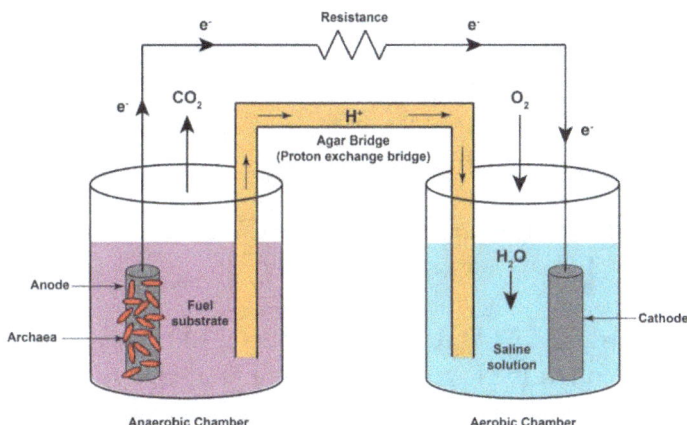

Fig. 1: Schematic outline of single unit MFC.

- **Electrodes**

Two 3D-graphite bars with the dimensions of 6*3*1 cm were used as the anode and cathode electrodes in both control and experimental units. Before use, both electrodes were immersed in 1 M HCl followed by 1 M NaOH for 1 h, and were subsequently stored in sterile distilled water **(Chen et al., 2015)**.

- The composition of the anolyte and catholyte solutions

The solution of the anolyte was consisting of 300 ml basal medium supplemented with 120 µM of phenolphthalein and inoculated with 10% of the archaeal strain. While the solution of the catholyte was 300 ml of normal saline (0.9%).

- Wire connections

The stainless steel wires with the diameter of 1 mm with 1000 Ω resistor were used for externally connecting the anode and cathode electrodes.

- The proton exchange bridge

A U-shape glass tube filled with 1.5% sterile agar was used to connect the anode and cathode chamber of each MFC unit in order to allow the passage of the protons.

- Power measurement

The produced current was calculated according to Ohm's law ($I = VR^{-1}$), where V is the voltage and R the applied resistor. The density of the current, j (mA m^{-2}), was calculated as $j = IS^{-1}$, where S is the surface area of the anode. The power density , P (mW m^{-2}), was calculated as $P = IVS^{-1}$ **(Logan et al., 2006)**.

Statistical optimization of anolyte composition using Plackett-Burman design (PBD)

A total number of 10 variables were tested for their relative significance according to **(Abdel-Fattah et al., 2007)** with some modifications. Some of the tested factors were related to the media components, while the others were physical parameters. Each variable was examined at the low lever (-1) and the high level (+1). All the examined factors were screened in 12 experimental trials. Each trial was tested in a chamber containing 300 ml of the culture media, and the response of each one was investigated through the measured voltage using a multimeter instrument.

Plackett-Burman design is based on a first order model:
$Y= \beta 0 + \sum \beta i \, xi.$

Where Y is the response (the measured voltage), β_0 is the model intercept, β_i is the variable calculate, and x_i represents the variable. The Pareto plot demonstrates the results of Plackett-Burman design since it illustrates the absolute relative significance of variables independent on their nature **(Plackett & Burman, 1946)**.

Verification experiment
A confirmatory optimization formula was examined depending on the results obtained from the most significant variables at their optimum levels as reported by the Plackett-Burman design.

(b) Stacked MFCs
Compared with the single cell MFC, two different designs named parallel and series MFC were tested for their overall power production **(Fig. 2)**. Each chamber of the three MFC design included a cylindrical graphite bar with the dimensions of 6*0.5 cm and a density of 0.51 g/cm, that has been extracted from waste batteries. All the applied conditions in this experiment was depending on the recommended optimum conditions of the Placket-Burman design.

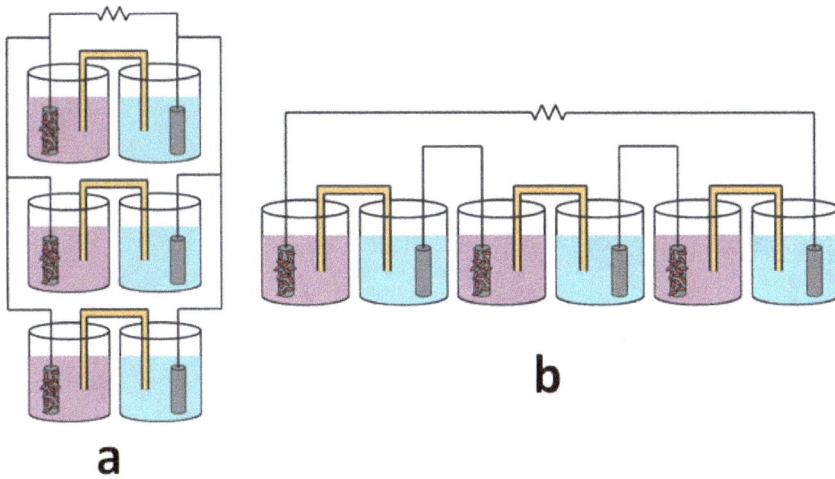

Fig. 2: Schematic outline of parallel stacked MFC (a), and series stacked MFC (b).

Results and discussion
Phenotypic characterization and growth curve of the halophilic archaeon *Natrialba* sp. GHMN55

The archaeon isolate *Natrialba* sp. GHMN55 (under accession number ac: MW794195) has been shown as a round orange-pigmented Gram negative bacteria **(Fig. 3a)**. Both of light and scanning electron micrographs showed the cells as cocci-shaped **(Fig. 3b and c)**. **Fig. 4** is illustrating the growth curve of the archaeon, where the lag phase was almost 4 days followed by 11 days' log phase. After spending 12 days in the stationary phase, the organism was entered the death phase with gradual decreasing in the optical density readings **(Hegazy et al., 2020)**.

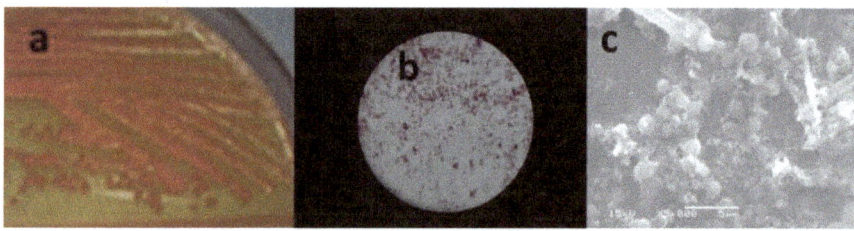

Fig. 3: Morphological and microscopic examinations of *Natrialba* sp. GHMN55.

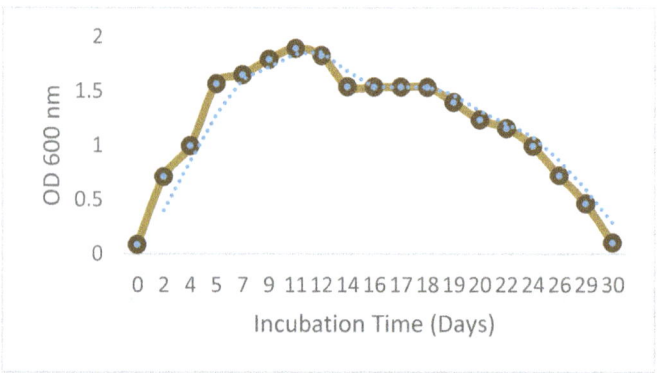

Fig. 4: Growth curve of *Natrialba* sp. GHMN55.

Determination of the polarization and power density of the single MFC

After 30 days of starting the experiment, both of the polarization and power density values of the control (organism lacking cell) and the treatment cells were investigated **(Fig. 5 and 6)**. It has been revealed that after 24 days of incubation, the maximum produced voltage in the treatment cell was 517 mV compared with 13 mV of the control cell **(Fig. 5)**. Moreover, at the same incubation time (24 days), the maximum power density of the treatment cell was recoded as 4949.7 mW/m^2 compared with 3.1 mW/m^2 of the control cell **(Fig. 6)**.

The maximum production of voltage or power density after 24 days could be attributed to the trend of the typical phases of the tested archaeal growth **(Boas et al., 2019, Simeon et al., 2020)**. As shown in **Fig. (4)**, the organism has started the decline phase as a result of the consumption of the organic matter after 20 days of incubation, which is highly matched with the obtained voltage and power density data.

On the other hand, both of voltage and power density values of the treatment cell were decreased to 190 mV and 668.5 mW/m^2 after 48 h of incubation (end of the experiment). The measurement drop of both parameters could be attributed to the depletion of the carbon source and would lead to a decrease in the microbial metabolism. It is predicted that the amendment of the cell with fresh carbon source could enhance the prolonged stability of the outcome power of the cell.

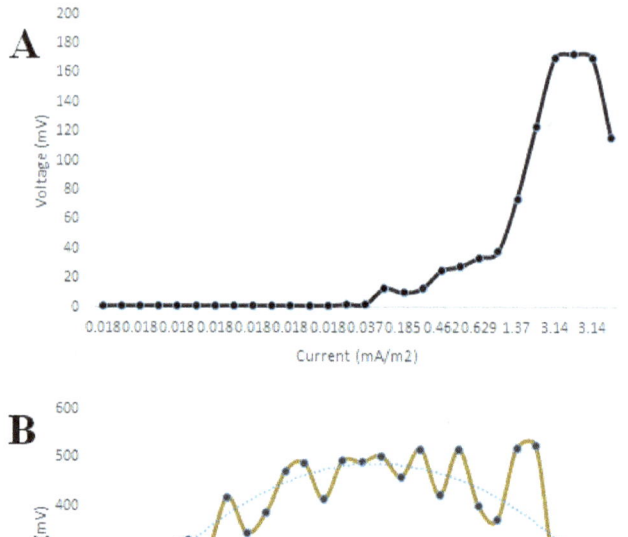

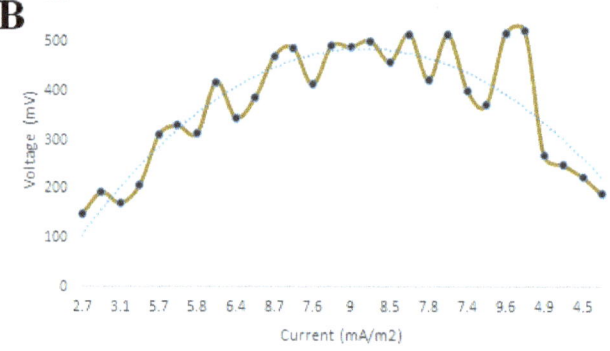

Fig. 5: Polarization curves of the designed MFC. A: the control cell, and B: the treatment cell.

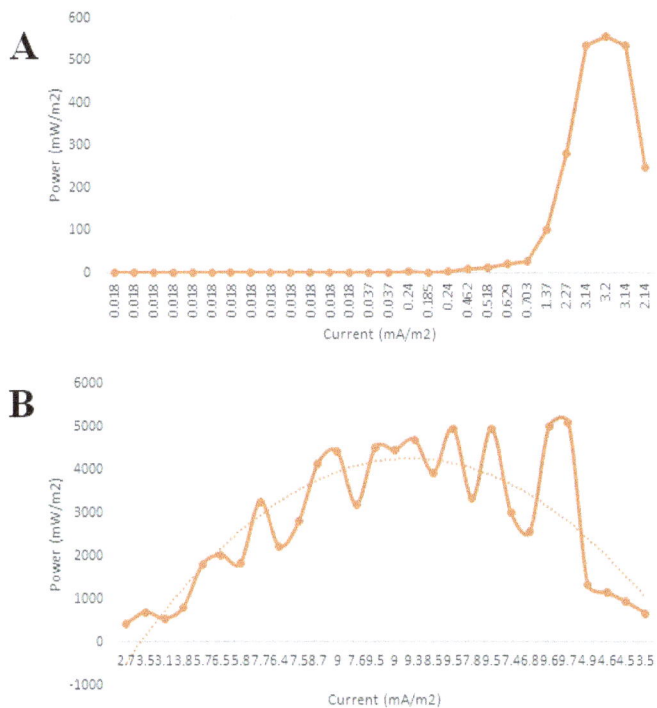

Fig. 6: Power density curves of the designed MFC. A: the control cell, and B: the treatment cell.

The composition of the anolyte medium has a very significant influence on the microbial activity in the fuel cell. This is where archaea act as catalysts for anaerobic oxidation of organics present in the anode chamber and can then generate electrons and protons, respectively, which are transferred to the cathode via wires and bridges. The composition of the anolyte depends on the type of the microorganisms used in the bioelectric generation process. Furthermore, the microbial performance of fuel cells is affected by the ionic strength of the anolyte solution. PBD was used to evaluate the importance of 10 culture factors including anolyte media composition and other microbial parameters of the fuel cell to optimize bioelectricity generation using the haloalkaliphilic archaea *Natrialba* sp. GHMN55 as casein, NaCl, pH, inoculum size, culture age, oxygen, magnet, intermediate, resistor value, and time. Based on PBD, each factor is tested at 2 levels: '-1' for low and '+1' for high. The matrix design of tested factors is presented in **Table (1)** and **Fig. (7)** next to the corresponding results for the tested factors.

Table 1: Optimization of anolyte solution in Microbial Fuel Cell using statistical experimental design (PBD).

Trials	X1 Casein (g/L)	X2 NaCl	X3 pH	X4 Inoculum Size (%)	X5 Culture Age (d)	X6 Oxygen	X7 Magnet	X8 Mediator PP (µM)	X9 Resistance (Ω)	X10 Time	Response (mV)
1	10(+)	15(-)	12(+)	5(-)	3(-)	Nil(-)	P(+)	120(+)	1000(+)	24(-)	0
2	10(+)	20(+)	10(-)	10(+)	3(-)	Nil(-)	Nil(-)	120(+)	1000(+)	48(+)	124
3	5(-)	20(+)	12(+)	5(-)	6(+)	Nil(-)	Nil(-)	Nil(-)	1000(+)	48(+)	5
4	10(+)	15(-)	12(+)	10(+)	3(-)	P(+)	Nil(-)	Nil(-)	100(-)	48(+)	0
5	10(+)	20(+)	10(-)	10(+)	6(+)	Nil(-)	P(+)	Nil(-)	100(-)	24(-)	178
6	10(+)	20(+)	12(+)	5(-)	6(+)	P(+)	Nil(-)	120(+)	100(-)	24(-)	68
7	5(-)	20(+)	12(+)	10(+)	3(-)	P(+)	P(+)	Nil(-)	1000(+)	24(-)	39

Trials	Casein (g/L) X1	NaCl X2	pH X3	Inoculum Size (%) X4	Culture Age (d) X5	Oxygen X6	Magnet X7	Mediator PP (µM) X8	Resistance (Ω) X9	Time X10	Response (mV)
8	5(-)	15(-)	12(+)	10(+)	6(+)	Nil(-)	P(+)	120(+)	100(-)	48(+)	0
9	5(-)	15(-)	10(-)	10(+)	6(+)	P(+)	Nil(-)	120(+)	1000(+)	24(-)	27
10	10(+)	15(-)	10(-)	5(-)	6(+)	P(+)	P(+)	Nil(-)	1000(+)	48(+)	200
11	5(-)	20(+)	10(-)	5(-)	3(-)	P(+)	P(+)	120(+)	100(-)	48(+)	0
12	5(-)	15(-)	10(-)	5(-)	3(-)	Nil(-)	Nil(-)	Nil(-)	100(-)	24(-)	67

Trials	X1 Casein (g/L)	X2 NaCl	X3 pH	X4 Inoculum Size (%)	X5 Culture Age (d)	X6 Oxygen	X7 Magnet	X8 Mediator PP (µM)	X9 Resistance (Ω)	X10 Time	Response (mV)
Coefficients	36	10	-40.3	2.3	20.6	-3.3	10.5	-22.5	6.8	-4.1	-
Main effect	72	20	-80.6	4.6	41.2	-6.6	21	-45	13.6	-8.2	-
p-value	0.374	0.748	0.341	0.938	0.547	0.912	0.737	0.520	0.823	0.890	-
t stat	1.5	0.416	-1.68	0.097	0.861	-0.138	0.437	-0.937	0.284	-0.173	-

Note: low level coded (-); high level coded (+) for the independent variables (X1-X10) presented between brackets are expressed in terms of g (w-/v%) or value. Phenolphthalene (PP), present (P), Not present (Nil).

Statistical analysis of the PBD

PBD includes a linear polynomial correlation equation model that illustrates the correlation between the 10 factors and the response as follows:

$$Y = 59 + 36 X_1 + 10X_2 - 40.33X_3 + 2.33 X_4 + 20.66 X_5 - 3.33 X_6 + 10.5X_7 - 22.5 X_8 + 6.83 X_9 - 4.16 X_{10}.$$

A large difference in the results of the Plackett-Burman experimental design was recorded, the maximum voltage value, 200 mV, was obtained in test number 10 after 48 h, while the minimum achievable was 0 mV, achieved in trials numbers 1, 4, 8, and 11. according to the main effect in **Fig. (7a)**, the most important factors affecting bioelectricity production were casein, age of inoculum, magnet, NaCl, resistance value, inoculum size, oxygen, time, intermediate material and pH in a decreasing order. By analyzing the regression coefficient R for 10 variables, it was found that casein, age of inoculation, magnet, NaCl, resistor value, and the size of inoculum had a positive effect on the bioelectricity production, which means that the high concentrations of these variables are close to optimal value. On the other hand, time, mediators and pH showed a negative effect on bioelectricity production, that is, low concentrations of these variables were close to optimal value, while the main effect was close to zero. This means that this factor has little or no effect on bioelectricity production.

The concentration of the carbon source affects the microbial growth rate and is directly proportional to the performance of the MFC and also based on the size of the inoculum **(Islam et al., 2018)**. Furthermore,

(**Rahimnejad et al., 2011**) reported that the performance of MFC was increased by increasing the concentration of glucose used by *Saccharomyces cerevisiae*. (**Li et al., 2018**) reported that many factors affect the performance of MFCs, such as physics (material and resistance of the electrodes), chemistry (oxidizing organic matter) and biology (type and age of bacteria used). (**Borole et al., 2008**) reported that MFC activity at low anodic pH increased the proton transfer rate and allowed more protons at the cathode surface, thereby increasing the potential for current generation.

The Pareto chart has been described as a very useful tool for presenting the most important impacts (**Fig. 8b**). It displays the significance of each variable and is a convenient way to check the results of the Plackett-Burman design. In this graph, the length of each variable bar on a normalized Pareto chart is proportional to the value of each variable for which the main effect or regression coefficient is calculated. The t-test for any variable effect tells us the probability of the calculated effect occurring by chance. Statistical confidence = $(1-p)*100$. Where $P = 0.05$ corresponds to a statistical confidence level of 95%. Therefore, any variable with statistical confidence close to or greater than 95% is considered significant. The accuracy of the model is guaranteed by the coefficient of determination (R2). If the calculated R2 value is close to 1.0, it is considered that there is a very strong correlation between the calculated data and the observed data; therefore, the current R2 value (0.87) reflects a very good fit between the calculated and observed responses, and ensures that this statistical model is reliable in predicting production. electricity (**El-Badan et al., 2020**).

The model's validation

According to the data obtained from the Plackett-Burman design, the following medium composition is predicted to be the optimum: NaCl, 200 g/l; Casamnoacids, 10 g/l; inoculums size, (10% (v/v)); pH, 10; inoculums age, 6 days; and external resistor, 1000 Ω; with the presence of magnetic bar attached to the anode, but with the absence of the mediator at anaerobic conditions with incubation time of 24 h.

The accuracy of the Plackett-Burman design was confirmed through this verification experiment. At the end of this experiment, the maximum output voltage of 231 mV was detected. This result decreased the needing time in power generation when compared with the results obtained by the basal conditions (**El-Badan et al., 2020**).

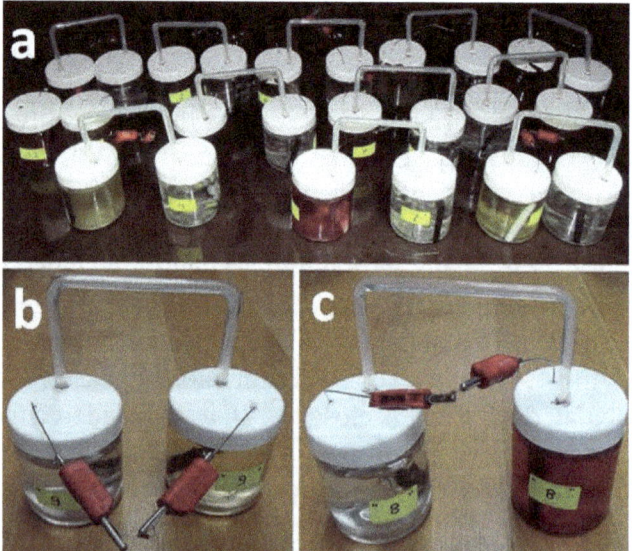

Fig. 7: The microbial fuel cell units of the Plackett-Burman design (a); single microbial fuel cell without mediator (b); and single microbial fuel cell with mediator (c).

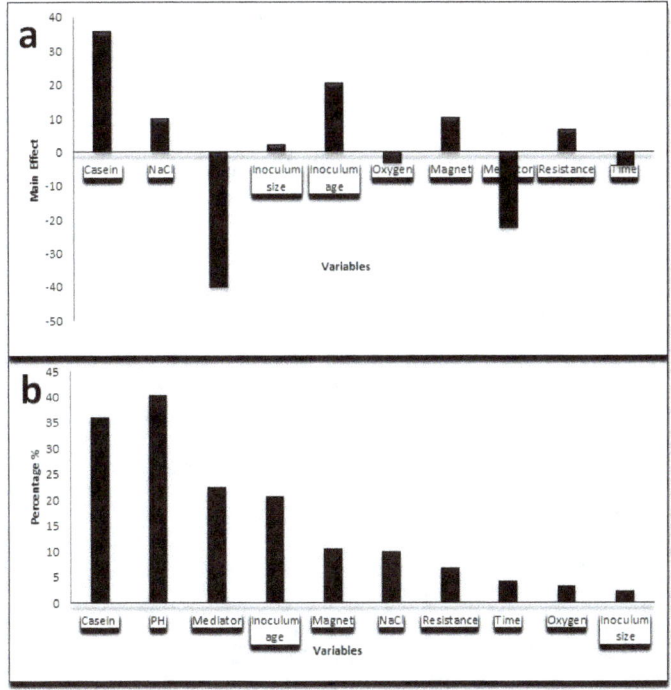

Fig. 8: (a) Main effect chart, (b) Pareto chart, rationalizing the effect of each factor on the voltage production by *Natrialba* sp. GHMN55.

SEM

Fig. (9) illustrated the SEM micrographs of 3 and 6 incubation days of the archaeal strain *Natrialba* sp. GHMN55 in addition to the anode surface with and without the archaeal strain. As can be shown in **Fig. (9a)**, the 3 days' incubation of the strain appeared as small round cells with lower cell density comparted with the 6 days of incubation that showed large round cells with higher cell density **(Fig. 9b)**.

On the other hand, the surface of the anode bar without the microbial cells appeared as coarse and un-smooth which can allow the microbial cells to get easily attached to this surface and easily fill the surface cavities using their ability to form biofilm **(Fig. 9c)**.

this assumption has been confirmed through the micrographs of the anode surface at the presence of the microbial cells, where a high cell density of the cells incorporated into the cavities of the graphite bar was investigated as can be shown in **Fig. (9d)**.

Fig. 9: SEM of free and electrode-coating *Natrialba* sp. GHMN55 after three and six days of incubation. (a): Free *Natrialba* sp. GHMN55 cells after 3 days of incubation, (b): Free *Natrialba* sp. GHMN55 cells after 6 days of incubation, (c): Archaeal cells-free graphite anodic electrode, and (d): Archaeal cells-bounded graphite anodic electrode.

Stacked MFCs

As the overall power produced by single MFC is low, many scientists tried to figure out multiple stacked MFCs such as series MFCs or parallel MFCs in order to increase the overall produced current or voltage **(Aelterman et al., 2006)**. In the current experiment, the voltage and power density of single, series, and parallel stacked MFCs were investigated. After 48 h of incubation, it has been shown that increasing the applied resistor from 90 to 50000 Ω resulted in gradual increasing of the produced voltage by all tested MFCs configurations (single, parallel, and series), as depicted in **Fig. (10)**. It has been shown that, after 48 h, the highest voltage of 27.5 mV was obtained by series units compared with 12.6 and 1.1 mV obtained from parallel and single MFCs, respectively. It has been observed that the power densities of the three units was also varied. As depicted in **Fig. (11)**, after 48 h of incubation, the maximum power density of 0.034 mW/m^2 was recorded by the parallel MFC when 10000 Ω resistor was applied. This value was reduced to 0.015 mW/m^2 when 50000 Ω was applied.

However, this behavior was remarkably differing in the single and series MFCs. They showed ascending pattern, where their power densities were recorded as 0.0001 and 0.075 mW/m^2 as the maximum power densities when 50000 Ω resistor was applied. From these results, it would be concluded that regarding the maximum power densities, the preferred designed MFCs are in the following order: series > parallel > single.

On the other hand, the behavior of the cells regarding the current density was the same as that of the power density. It has been shown that, the maximum power densities of 0.0001, 0.001, and 0.002 mA/m^2 were recorded by the single, parallel, and series MFCs, respectively, which indicate that the parallel design is not the preferred one for the current study. We are not in agreement with **(Vilajeliu-Pons et al., 2017)** who recorded that, regarding the produced power, the parallel electrical configuration was preferred than the series configuration. It has been reported that series connection may suffer from losses in the contact voltage or reversal of the voltage, while parallel connections are suffering from increases in the internal losses which reduces the total power production **(Oh & Logan, 2007).** We could attribute the lower power densities of the current experiment compared with the single cell MFC performed at the begging of the current work to the size area of the electrode which plays an important and significant role in the overall produced power **(Shirpay, 2021).**

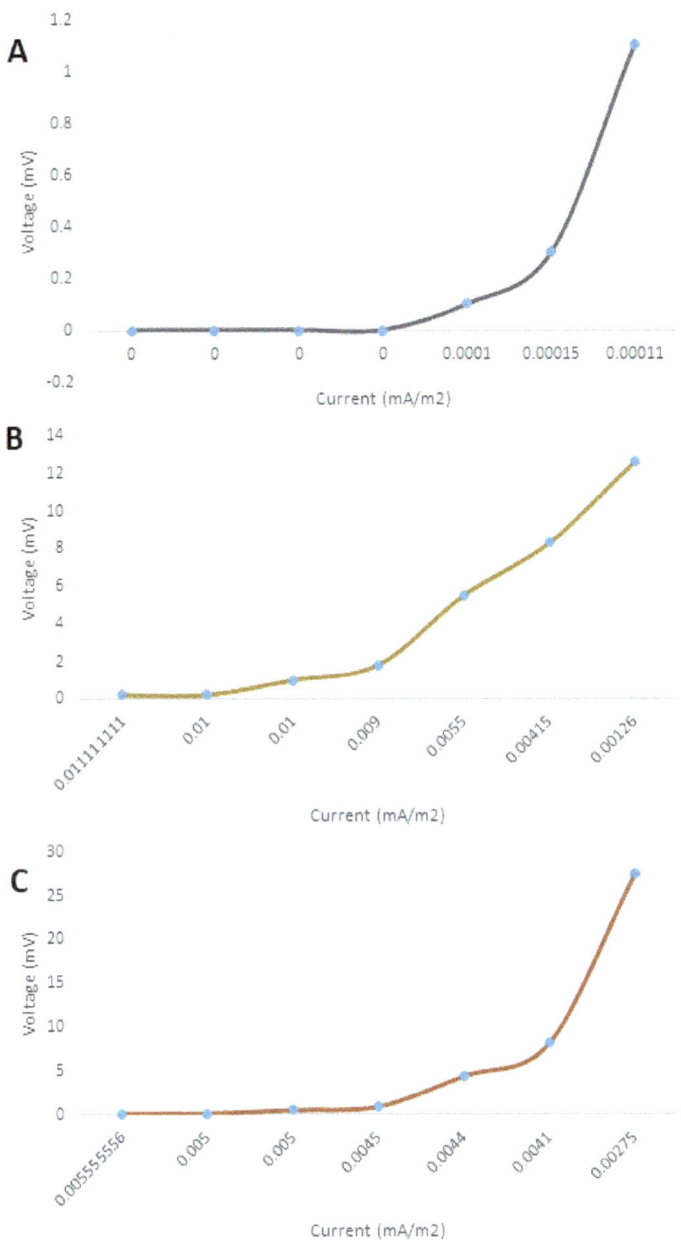

Fig. 10: Polarization curves of the stacked MFC. A: Single unit, B: Parallel units, and C: Series units.

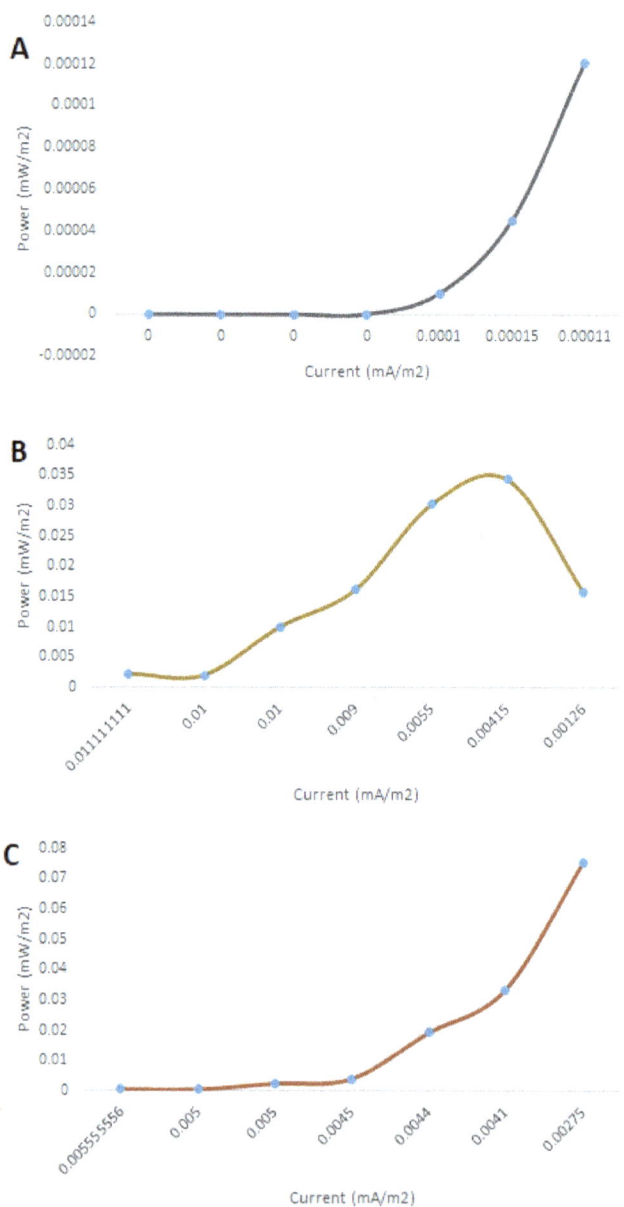

Fig. 11: Power density curves of the stacked MFC. A: Single unit, B: Parallel units, and C: Series units.

Conclusion

The production of bioelectricity by the isolate *Natrialba* sp. GHMN55 that is considered as alkaliphilic and halophilic archaeon has been investigated. After confirmed production of electricity by this microorganism, the Plackett-Burman experimental design has been applied in order to reduce the time required for the power production through the tested microbial fuel cell. Some parameters such as casein, magnet, inoculum age, resistor value, NaCl, and inoculum size Showed high power generation when used in high concentrations, while other parameters such as mediator, time, and pH are preferred to be used in lower values. The results of the stacked series and parallel MFC compared with single units showed that both of parallel and series MFC units are preferred for maximum power production than single cells, while the series design is the ideal one for the current study.

References

Abdel-Fattah Y, Soliman N, Berekaa M, 2007. Application of Box-Behnken design for optimization of poly-γ-glutamic acid production by Bacillus licheniformis SAB-26. *Res. J. Microbiol* **2**, 664-70.

Aelterman P, Rabaey K, Pham HT, Boon N, Verstraete W, 2006. Continuous electricity generation at high voltages and currents using stacked microbial fuel cells. *Environmental science & technology* **40**, 3388-94.

Bishoge OK, Zhang L, Mushi WG, 2019. The potential renewable energy for sustainable development in Tanzania: A review. *Clean Technologies* **1**, 70-88.

Boas JV, Oliveira V, Marcon L, Simões M, Pinto A, 2019. Optimization of a single chamber microbial fuel cell using Lactobacillus pentosus: Influence of design and operating parameters. *Science of The Total Environment* **648**, 263-70.

Bond DR, Lovley DR, 2003. Electricity production by Geobacter sulfurreducens attached to electrodes. *Applied and environmental microbiology* **69**, 1548-55.

Cao TN-D, Chen S-S, Ray SS, Le HQ, Chang H-M, 2019. Application of microbial fuel cell in wastewater treatment and simultaneous bioelectricity generation. In. *Water and Wastewater Treatment Technologies.* Springer, 501-26.

Chen Z, Li K, Zhang P, Pu L, Zhang X, Fu Z, 2015. The performance of activated carbon treated with H3PO4 at 80° C in the air-cathode microbial fuel cell. *Chemical Engineering Journal* **259**, 820-6.

El-Badan ES, A Khaled M, M Ghanem K, 2020. Optimization of anolyte solution in Microbial Fuel Cell using statistical experimental design. *Egyptian Journal of Aquatic Biology and Fisheries* **24**, 173-87.

Hegazy GE, Abu-Serie MM, Abo-Elela GM, *et al.*, 2020. In vitro dual (anticancer and antiviral) activity of the carotenoids produced by haloalkaliphilic archaeon Natrialba sp. M6. *Scientific reports* **10**, 1-14.

Hegazy GE, Taha TH, Abdel-Fattah YR, 2022. Investigation of the optimum conditions for electricity generation by haloalkaliphilic archaeon Natrialba sp. GHMN55 using the Plackett–Burman design: single and stacked MFCs. *Microbial Cell Factories* **21**, 1-14.

Kaur A, Ibrahim S, Pickett CJ, *et al.*, 2014. Anode modification to improve the performance of a microbial fuel cell volatile fatty acid biosensor. *Sensors and Actuators B: Chemical* **201**, 266-73.

Li B, Zhou J, Zhou X, *et al.*, 2014. Surface modification of microbial fuel cells anodes:

approaches to practical design. *Electrochimica Acta* **134**, 116-26.

Logan BE, Hamelers B, Rozendal R, *et al.*, 2006. Microbial fuel cells: methodology and technology. *Environmental science & technology* **40**, 5181-92.

Oh S-E, Logan BE, 2007. Voltage reversal during microbial fuel cell stack operation. *Journal of Power Sources* **167**, 11-7.

Plackett RL, Burman JP, 1946. The design of optimum multifactorial experiments. *Biometrika* **33**, 305-25.

Shirpay A, 2021. Effects of electrode size on the power generation of the microbial fuel cell by Saccharomyces cerevisiae. *Ionics* **27**, 3967-73.

Simeon MI, Asoiro FU, Aliyu M, Raji OA, Freitag R, 2020. Polarization and power density trends of a soil-based microbial fuel cell treated with human urine. *International Journal of Energy Research* **44**, 5968-76.

Ulusoy I, Dimoglo A, 2018. Electricity generation in microbial fuel cell systems with Thiobacillus ferrooxidans as the cathode microorganism. *international journal of hydrogen energy* **43**, 1171-8.

Vilajeliu-Pons A, Puig S, Salcedo-Dávila I, Balaguer M, Colprim J, 2017. Long-term assessment of six-stacked scaled-up MFCs treating swine manure with different electrode materials. *Environmental Science: Water Research & Technology* **3**, 947-59.

Xu G, Zheng X, Lu Y, *et al.*, 2019. Development of microbial community within the cathodic biofilm of single-chamber air-cathode microbial fuel cell. *Science of the total environment* **665**, 641-8.

www.ingramcontent.com/pod-product-compliance
Lightning Source LLC
Chambersburg PA
CBHW072209290526
45794CB00004B/1708